Property of

Mark Williams
Handee-Pak, Inc.

QUESTIONS AND ANSWERS FOR ELECTRICIANS EXAMINATIONS

by Edwin P. Anderson

Revised and updated
by Roland E. Palmquist

THEODORE AUDEL & CO.
a division of
HOWARD W. SAMS & CO., INC.
4300 West 62nd Street
Indianapolis, Indiana 46268

FIFTH EDITION

SECOND PRINTING—1976

Copyright © 1975 by Howard W. Sams & Co., Inc., Indianapolis, Indiana 46268. Printed in the United States of America.

All rights reserved. Reproduction or use, without express permission, of editorial or pictorial content, in any manner, is prohibited. No patent liability is assumed with respect to the use of the information contained herein. While every precaution has been taken in the preparation of this book, the publisher assumes no responsibility for errors or omissions. Neither is any liability assumed for damages resulting from the use of the information contained herein.

International Standard Book Number: 0-672-23225-1
Library of Congress Catalog Card Number: 70-189288

Foreword

This book is offered as a guide for anyone preparing for the various electricians' license examinations—Apprentice, Journeyman, or Master. These examinations are given periodically by local licensing authorities, either municipal, county, state, or any other agency having legal jurisdiction over the licensing and inspection of work done by electricians.

To ply his trade, an electrician is required by law in most localities to secure a license from the enforcing authority in his area. This book supplies the license applicant with the required theoretical knowledge and a thorough understanding of the definitions, specifications, and regulations of the 1975 National Electrical Code on which the majority of these examinations is based.

The National Electrical Code has been prepared by the National Fire Protection Association with the assistance of men in all phases of the electrical industry. Anyone in the field of electricity may contribute to the Code after first proving the need for change, thereby making this document both flexible and democratic.

Numerous examples in the form of questions and answers are presented thoughout this book, thereby enabling the license applicant to gain a complete knowledge of the types of questions asked in an electricians' examination. The tremendous importance of careful study in order to master the fundamental principles underlying each question and answer should be thoroughly impressed on all candidates for licenses in the various grades. Only through this process can the license applicant prepare himself to solve any new or similar problem on his examination.

The interpretations of the NEC are those of the author and are not necessarily the official observations of the National Electrical Code Committee.

ROLAND E. PALMQUIST

Contents

ELECTRICAL SYMBOLS 9

CHAPTER 1
REVIEW DEFINITIONS 45

CHAPTER 2
OHM'S LAW AND OTHER ELECTRICAL FORMULAS 53

CHAPTER 3
POWER AND POWER FACTOR 85

CHAPTER 4
LIGHTING 99

CHAPTER 5
BRANCH CIRCUITS AND FEEDERS 107

CHAPTER 6
TRANSFORMER PRINCIPLES AND CONNECTIONS 125

CHAPTER 7
WIRING DESIGN AND PROTECTION 143

CHAPTER 8
WIRING METHODS AND MATERIALS 165

CHAPTER 9
BATTERIES AND RECTIFICATION 183

CHAPTER 10
VOLTAGE GENERATION 187

CHAPTER 11
EQUIPMENT FOR GENERAL USE 193

CHAPTER 12
MOTORS 211

CHAPTER 13
MOTOR CONTROLS 227

CHAPTER 14
SPECIAL OCCUPANCIES AND HAZARDOUS LOCATIONS 247

CHAPTER 15
GROUNDING AND GROUND TESTING 265

INDEX 278

ELECTRICAL SYMBOLS

ASA policy requires that the same symbol not be included in two or more Standards. The reason for this is that if the same symbol was included in two or more Standards, when a symbol included in one Standard was revised, it might not be so revised in the other Standard at the same time leading to confusion as to which was the proper symbol to use.

2. Symbols falling into the above category include, but are not limited to, those shown below. The reference numbers are the American Standard Y32.2 item numbers.

(MOT)   46.3 Electric motor

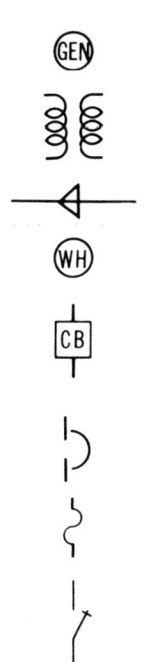

(GEN)   46.2 Electric generator

     86.1 Power transformer

     82.1 Pothead (cable termination)

(WH)   48   Electric watthour meter

[CB]   12.2 Circuit element, e.g., circuit breaker

     11.1 Circuit breaker

     36   Fusible element

     76.3 Single-throw knife switch

     76.2 Double-throw knife switch

     13.1 Ground

ELECTRICAL SYMBOLS

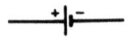

——+|–——   7   Battery

LIST OF SYMBOLS

1.0 Lighting Outlets

Ceiling   **Wall**

1.1 Surface or pendant incandescent mercury vapor or similar lamp fixture

1.2 Recessed incandescent mercury vapor or similar lamp fixture

1.3 Surface or pendant individual fluorescent fixture

1.4 Recessed individual fluorescent fixture

1.5 Surface or pendant continuous-row fluorescent fixture

1.6 *Recessed continuous-row fluorescent fixture

1.7 **Bare-lamp fluorescent strip

1.8 Surface or pendant exit light

1.9 Recessed exit light

ELECTRICAL SYMBOLS

1.10 Blanked outlet

1.11 Junction box

1.12 Outlet controlled by low-voltage switching when relay is installed in outlet box

2.0 Receptacle Outlets

Where all or a majority of receptables in an installation are to be of the grounding type, the upper-case letter abbreviated notation may be omitted and the types of receptacles required noted in the drawing list of symbols and/or in the specifications. When this is done, any non-grounding receptacles may be so identified by notation at the outlet location.

Where weather proof, explosion proof or other specific types of devices are to be required, use the type of upper-case subscript letters referred to under Section 0.2 item a-2 of this Standard. For example, weather proof single or duplex receptacles would have the upper-case subscript letters noted alongside of the symbol.

Ungrounded   **Grounding**

2.1 Single receptacle outlet

2.2 Duplex receptacle outlet

2.3 Triplex receptacle outlet

2.4 Quadruplex receptacle outlet

2.5 Duplex receptacle outlet — split wired

*In the case of combination continuous-row fluorescent and incandescent spotlights, use combinations of the above standard symbols.
**In the case of continuous-row bare-lamp flourescent strip above an area-wide diffusing means, show each fixture run, using the standard symbol; indicate area of diffusing means and type by light shading and/or drawing notation.

ELECTRICAL SYMBOLS

Ungrounded	Grounding		
		2.6	Triplex receptacle outlet — split wired
		2.7	*Single special-purpose receptacle outlet
		2.8	*Duplex special-purpose receptacle outlet
R	RG	2.9	Range outlet
DW	G DW	2.10	Special-purpose connection or provision for connection. Use subscript letters to indicate function (DW — dishwasher; CD — clothes dryer, etc.)
X″	G X″	2.11	Multi-outlet assembly. (Extend arrows to limit of installation. Use appropriate symbol to indicate type of outlet. Also indicate spacing of outlets as x inches.)
F	C G	2.12	Clock Hanger Receptacle
C	F G	2.13	Fan Hanger Receptacle
	G	2.14	Floor Single Receptacle Outlet
	G	2.15	Floor Duplex Receptacle Outlet

*Use numeral or letter either within the symbol or as a subscript alongside the symbol keyed to explanation in the drawing list of symbols to indicate type of receptacle or usage.

ELECTRICAL SYMBOLS

Ungrounded **Grounding**

2.16 *Floor Special-Purpose Outlet

2.17 Floor Telephone Outlet—Public

2.18 Floor Telephone Outlet—Private

Not a part of the Standard: Example of the use of several floor outlet symbols to identify a 2, 3, or more gang floor outlet

2.19 Underfloor Duct and Junction Box for Triple, Double or Single Duct System as indicated by the number of parallel lines

Not a part of the Standard: Example of use of various symbols to identify location of different types of outlets or connections for underfloor duct or cellular floor systems

2.20 Cellular Floor Header Duct

3.0 Switch Outlets

S   3.1 Single-pole switch

---

*Use numeral keyed to explanation in drawing list of symbols to indicate usage.

11

Electrical Symbols

S_2 3.2 Double-pole switch

S_3 3.3 Three-way switch

S_4 3.4 Four-way switch

S_K 3.5 Key-operated switch

S_P 3.6 Switch and pilot lamp

S_L 3.7 Switch for low-voltage switching system

S_{LM} 3.8 Master switch for low-voltage switching system

—⊖S 3.9 Switch and single receptacle

=⊖S 3.10 Switch and double receptacle

S_D 3.11 Door switch

S_T 3.12 Time switch

S_{CB} 3.13 Circuit breaker switch

S_{MC} 3.14 Momentary contact switch or pushbutton for other than signalling system

ELECTRICAL SYMBOLS

3.15 Ceiling pull switch

SIGNALLING SYSTEM OUTLETS

4.0 Institutional, Commercial, and Industrial Occupancies

Basic Symbol	Examples of Individual Item Identification (Not a part of the standard)		

4.1 **I. Nurse Call System Devices (any type)**

Nurses' Annunciator (can add a number after it as ─⊕─① 24 to indicate number of lamps)

Call station, single cord, pilot light

Call station, double cord, microphone-speaker

Corridor dome light, 1 lamp

Transformer

Any other item on same system—use numbers as required.

4.2 **II. Paging System Devices (any type)**

Keyboard

13

ELECTRICAL SYMBOLS

Basic Symbol	Examples of Individual Item Identification (Not a part of the standard)	
	─┤◇2	*Flush annunciator*
	─┤◇3	*2-Face annunciator*
	─┤◇4	*Any other item on same system—use numbers as required.*
─┤☐		**4.3 III. Fire Alarm System Devices (any type) including Smoke and Sprinkler Alarm Devices**
	─┤☐1	*Control panel*
	─┤☐2	*Station*
	─┤☐3	*10″ Gong*
	─┤☐4	*Pre-signal chime*
	─┤☐5	*Any other item on same system—use numbers as required.*

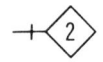

ELECTRICAL SYMBOLS

Basic Symbol   **Examples of Individual Item Identification (Not a part of the standard)**

4.4   **IV. Staff Register System Devices (any type)**

       ①   *Phone operators' register*

       ②   *Entrance register—flush*

       ③   *Staff room register*

       ④   *Transformer*

       ⑤   *Any other item on same system—use numbers as required.*

4.5   **V. Electric Clock System Devices (any type)**

       ⟨1⟩   *Master clock*

       ⟨2⟩   *12" Secondary—flush*

Electrical Symbols

Basic Symbol	Examples of Individual Item Identification (Not a part of the standard)	
	⊢⟨3⟩	12" Double dial—wall mounted
	⊢⟨4⟩	18" Skeleton dial
	⊢⟨5⟩	Any other item on same system—use numbers as required.
⊢◁		4.6 **VI. Public Telephone System Devices**
	⊢◁1	Switchboard
	⊢◁2	Desk phone
	⊢◁3	Any other item on same system—use numbers as required.
⊢◀		4.7 **VII. Private Telephone System Devices (any type)**
	⊢◀1	Switchboard
	⊢◀2	Wall phone
	⊢◀3	Any other item on same system—use numbers as required.

ELECTRICAL SYMBOLS

Basic Symbol	Examples of Individual Item Identification (Not a part of the standard)		
─┤△		4.8	**VIII. Watchman System Devices (any type)**
	─┤①		*Central station*
	─┤②		*Key station*
	─┤③		*Any other item on same system—use numbers as required.*
─┤◁		4.9	**IX. Sound System**
	─┤◁1		*Amplifier*
	─┤◁2		*Microphone*
	─┤◁3		*Interior speaker*
	─┤◁4		*Exterior speaker*
	─┤◁5		*Any other item on same system—use numbers as required.*
─┤▢		4.10	**X. Other Signal System Devices**
	─┤⑪		*Buzzer*
	─┤⑫		*Bell*

17

ELECTRICAL SYMBOLS

Basic Symbol   **Examples of Individual Item Identification** (Not a part of the standard)

                ─⊕③   *Pushbutton*

                ─⊕④   *Annunciator*

                ─⊕⑤   *Any other item on same system—use numbers as required.*

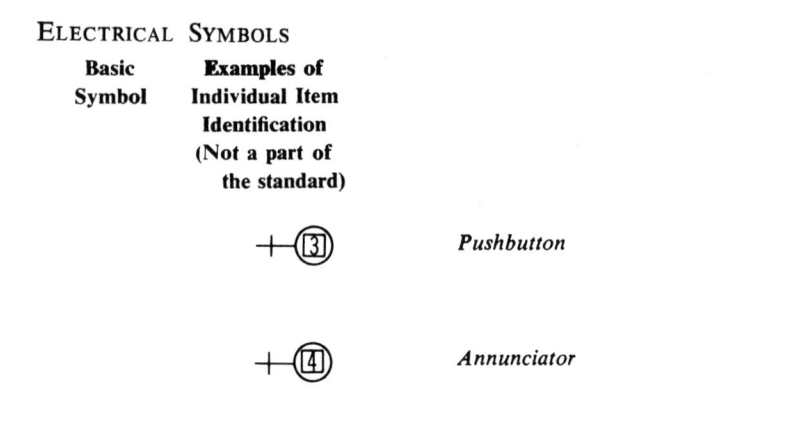

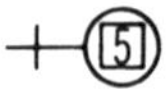

SIGNALLING SYSTEM OUTLETS

5.0 Residential Occupancies

Signalling system symbols for use in identifying standardized residential-type signal system items on residential drawings where a descriptive symbol list is not included on the drawing. When other signal system items are to be identified, use the above basic symbols for such items together with a descriptive symbol list.

     ▪   *5.1*   *Pushbutton*

     ◣   *5.2*   *Buzzer*

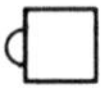

     ⌐▢   *5.3*   *Bell*

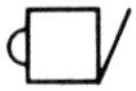

     ⌐▢/   *5.4*   *Combination bell-buzzer*

     [CH]   *5.5*   *Chime*

     ◇─   *5.6*   *Annunciator*

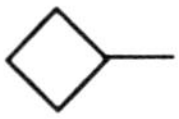

     [D]   *5.7*   *Electric door opener*

ELECTRICAL SYMBOLS

| M | 5.8 Maid's signal plug |

| ☐ | 5.9 Interconnection box |

| BT | 5.10 Bell-ringing transformer |

| ▶| | 5.11 Outside telephone |

| ▷| | 5.12 Interconnecting telephone |

| R | 5.13 Radio outlet |

| TV | 5.14 Television outlet |

6.0 Panelboards, Switchboards and Related Equipment

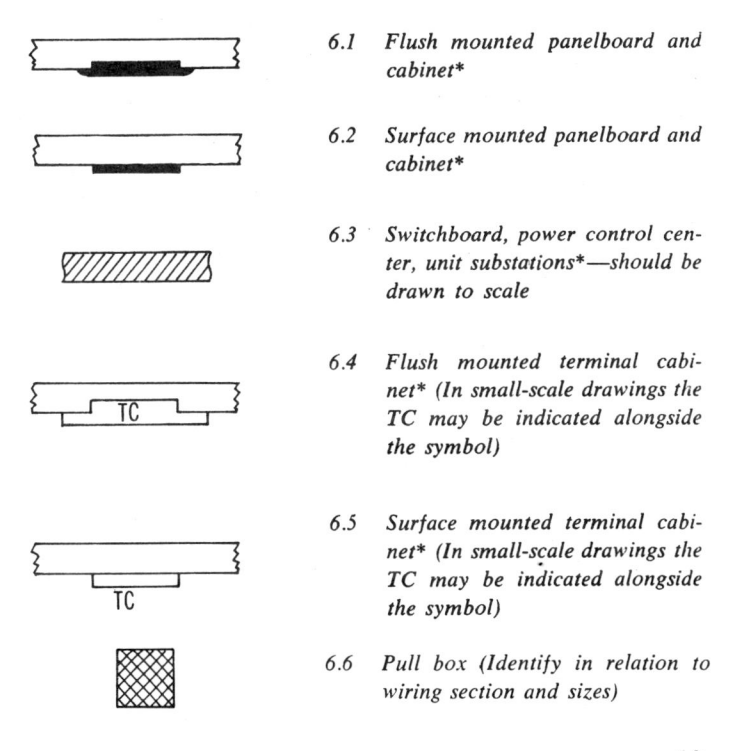

6.1 Flush mounted panelboard and cabinet*

6.2 Surface mounted panelboard and cabinet*

6.3 Switchboard, power control center, unit substations*—should be drawn to scale

6.4 Flush mounted terminal cabinet* (In small-scale drawings the TC may be indicated alongside the symbol)

6.5 Surface mounted terminal cabinet* (In small-scale drawings the TC may be indicated alongside the symbol)

6.6 Pull box (Identify in relation to wiring section and sizes)

19

ELECTRICAL SYMBOLS

6.7 Motor or other power controller*

6.8 Externally operated disconnection switch*

6.9 Combination controller and disconnection means*

7.0 Bus Ducts and Wireways

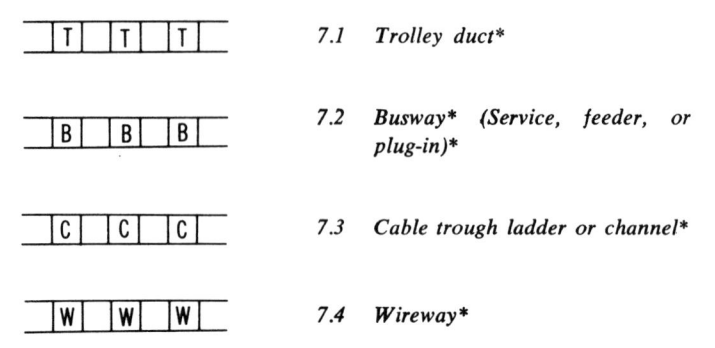

7.1 Trolley duct*

7.2 Busway* (Service, feeder, or plug-in)*

7.3 Cable trough ladder or channel*

7.4 Wireway*

8.0 Remote Control Stations for Motors or other Equipment*

8.1 Pushbutton station

8.2 Float switch—mechanical

8.3 Limit switch—mechanical

*Identify By Notation or Schedule.

Electrical Symbols

[P]	8.4 Pneumatic switch—mechanical
◨←	8.5 Electric eye—beam source
◨	8.6 Electric eye—relay
—(T)	8.7 Thermostat

9.0 Circuiting Wiring method identification by notation on drawing or in specification

───────	9.1 Wiring concealed in ceiling or wall
— — — — —	9.2 Wiring concealed in floor
- - - - - - - -	9.3 Wiring exposed

Note: Use heavy-weight line to identify service and feeders. Indicate empty conduit by notation CO (conduit only)

       2   1
──────→

9.4 Branch circuit home run to panel board. Number of arrows indicates number of circuits. (A numeral at each arrow may be used to identify circuit number.) Note: Any circuit without further identification indicates two-wire circuit. For a greater number of wires, indicate with cross lines, e.g.:

3 wires;
──/*/*/──

4 wires, etc.
──/*/*/*/──

Unless indicated otherwise, the wire size of the circuit is the minimum size required by the specification.

Electrical Symbols

Identify different functions of wiring system, e.g., signalling system by notation or other means.

———o   *9.5   Wiring turned up*

———•   *9.6   Wiring turned down*

10.0 Electric Distribution or Lighting System, Underground

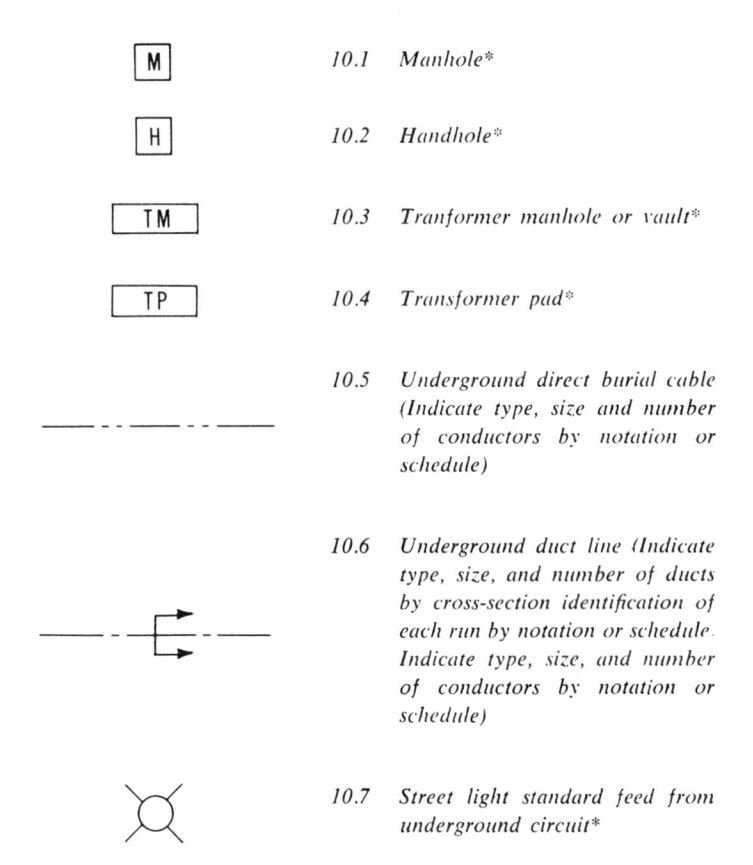

M	*10.1 Manhole**
H	*10.2 Handhole**
TM	*10.3 Tranformer manhole or vault**
TP	*10.4 Transformer pad**

10.5   Underground direct burial cable (Indicate type, size and number of conductors by notation or schedule)

10.6   Underground duct line (Indicate type, size, and number of ducts by cross-section identification of each run by notation or schedule. Indicate type, size, and number of conductors by notation or schedule)

10.7   Street light standard feed from underground circuit*

---

*Identify By Notation or Schedule.

ELECTRICAL SYMBOLS

11.0 Electric Distribution or Lighting System Aerial

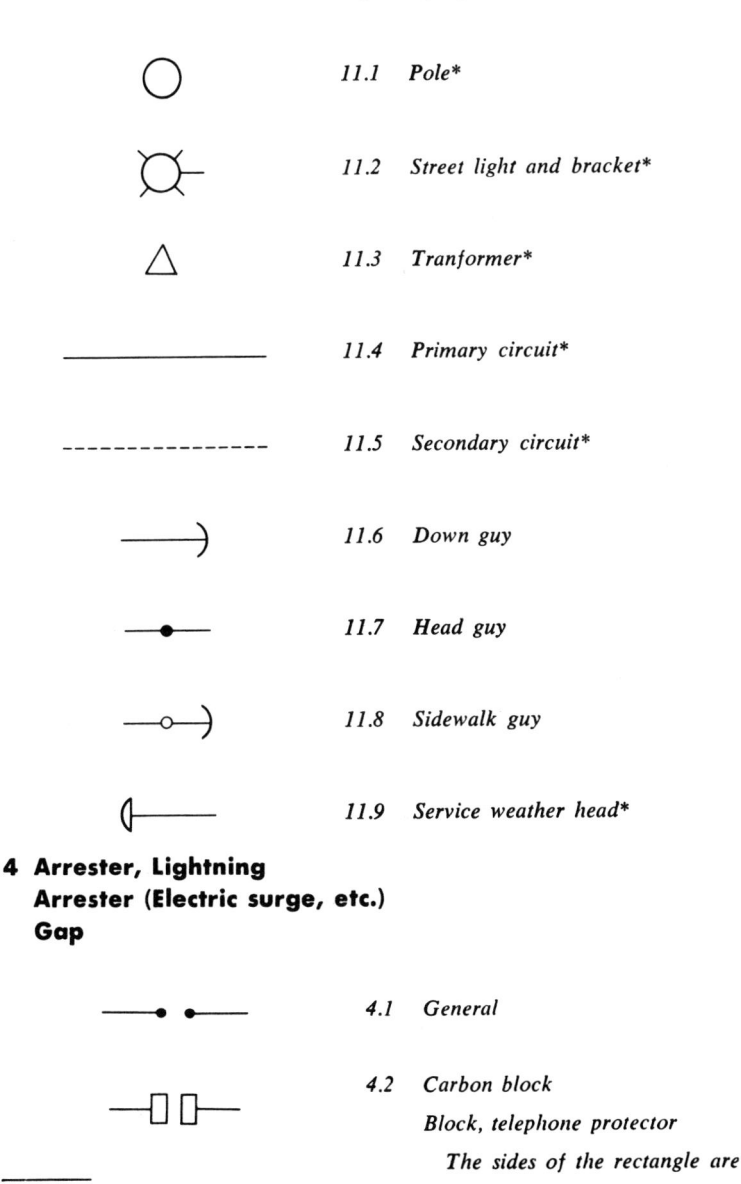

○	11.1	Pole*
⊗	11.2	Street light and bracket*
△	11.3	Tranformer*
———	11.4	Primary circuit*
- - - - -	11.5	Secondary circuit*
——)	11.6	Down guy
—•—	11.7	Head guy
—o—)	11.8	Sidewalk guy
(|—	11.9	Service weather head*

4 Arrester, Lightning Arrester (Electric surge, etc.) Gap

—• •—	4.1	General
—[] []—	4.2	Carbon block
		Block, telephone protector
		The sides of the rectangle are

*Identify By Notation or Schedule.

Electrical Symbols

        to be approximately in the ratio of 1 to 2 and the space between rectangles shall be approximately equal to the width of a rectangle.

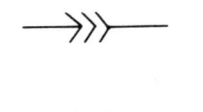

4.3   Electrolytic or aluminum cell
      This symbol is not composed of arrowheads.

4.4   Horn gap

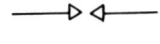

4.5   Protective gap
      These triangles shall not be filled.

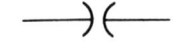

4.6   Sphere gap

4.7   Valve or film element

4.8   Multigap, general

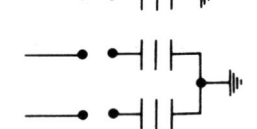

4.9   Application: gap plus valve plus ground, 2 pole

7 Battery

The long line is always positive, but polarity may be indicated in addition. Example:

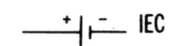

7.1   Generalized direct-current source

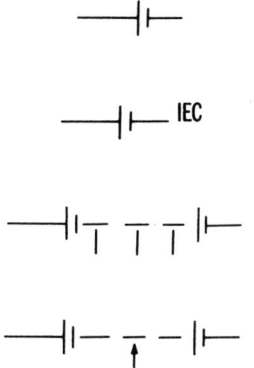

ELECTRICAL SYMBOLS

7.2   One cell

— | |— IEC   7.3   Multicell

7.3.1   Multicell battery with 3 taps

7.3.2   Multicell battery with adjustable tap

11 Circuit Breakers

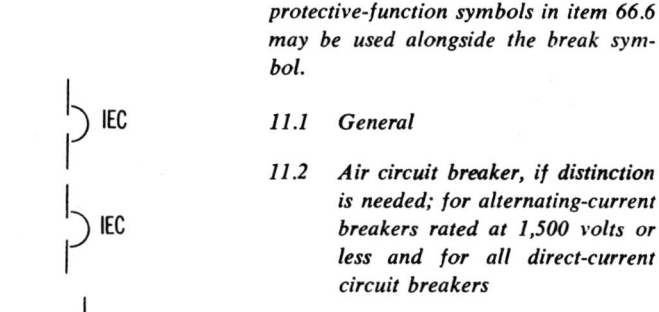

If it is desired to show the condition causing the breaker to trip, the relay-protective-function symbols in item 66.6 may be used alongside the break symbol.

11.1   General

11.2   Air circuit breaker, if distinction is needed; for alternating-current breakers rated at 1,500 volts or less and for all direct-current circuit breakers

11.2.1   Network protector

11.3   Circuit breaker, other than covered by item 11.2. The symbol in the right column is for a 3-pole breaker.

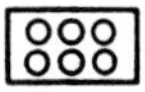

See note 11.3A

11.3.1   On a connection or wiring diagram, a 3-pole single-throw circuit breaker (with terminals shown) may be drawn as shown.

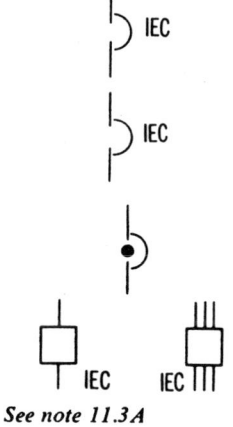

See note 11.3A

25

ELECTRICAL SYMBOLS

11.4 Applications

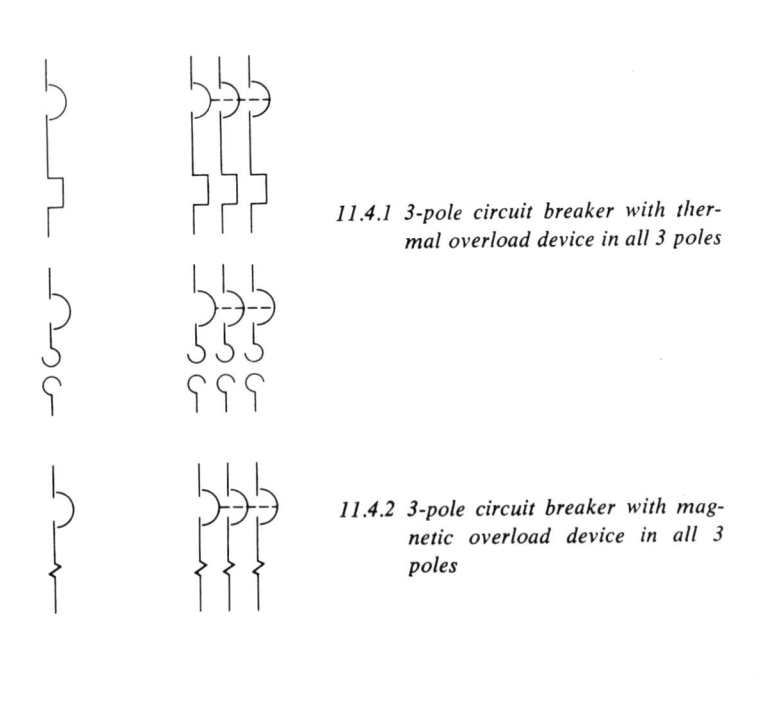

11.4.1 3-pole circuit breaker with thermal overload device in all 3 poles

11.4.2 3-pole circuit breaker with magnetic overload device in all 3 poles

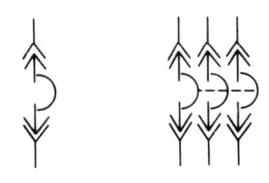

11.4.3 3-pole circuit breaker, drawout type

13 Circuit Return

13.1 Ground

Note 11.3A—On a power diagram, the symbol may be used without other identification. On a composite drawing where confusion with the general circuit element symbol (item 12) may result, add the identifying letters CB inside or adjacent to the square.

26

ELECTRICAL SYMBOLS

(A) *A direct conducting connection to the earth or body of water that is a part thereof.*

(B) *A conducting connection to a structure that serves a function similar to that of an earth ground (that is, a structure such as a frame of an air, space, or land vehicle that is not conductively connected to earth).*

13.2 Chassis or frame connection
    A conducting connection to a chassis or frame of a unit. The chassis or frame may be at a substantial potential with respect to the earth or structure in which this chassis or frame is mounted.

13.3 Common connections
    Conducting connections made to one another. All like-designated points are connected.
    *The asterisk is not a part of the symbol. Identifying values, letters, numbers, or marks shall replace the asterisk.

15 Coil, Magnetic Blowout ‡

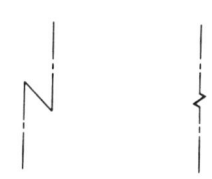

---

‡The broken line -——— indicates where line connection to a symbol is made and is not a part of the symbol.

27

ELECTRICAL SYMBOLS
23 Contact, Electrical

For build-ups or forms using electrical contacts, see applications under CONNECTOR (item 18), RELAY (item 66), SWITCH (item 76). See DRAFTING PRACTICES (item 0.4.6).

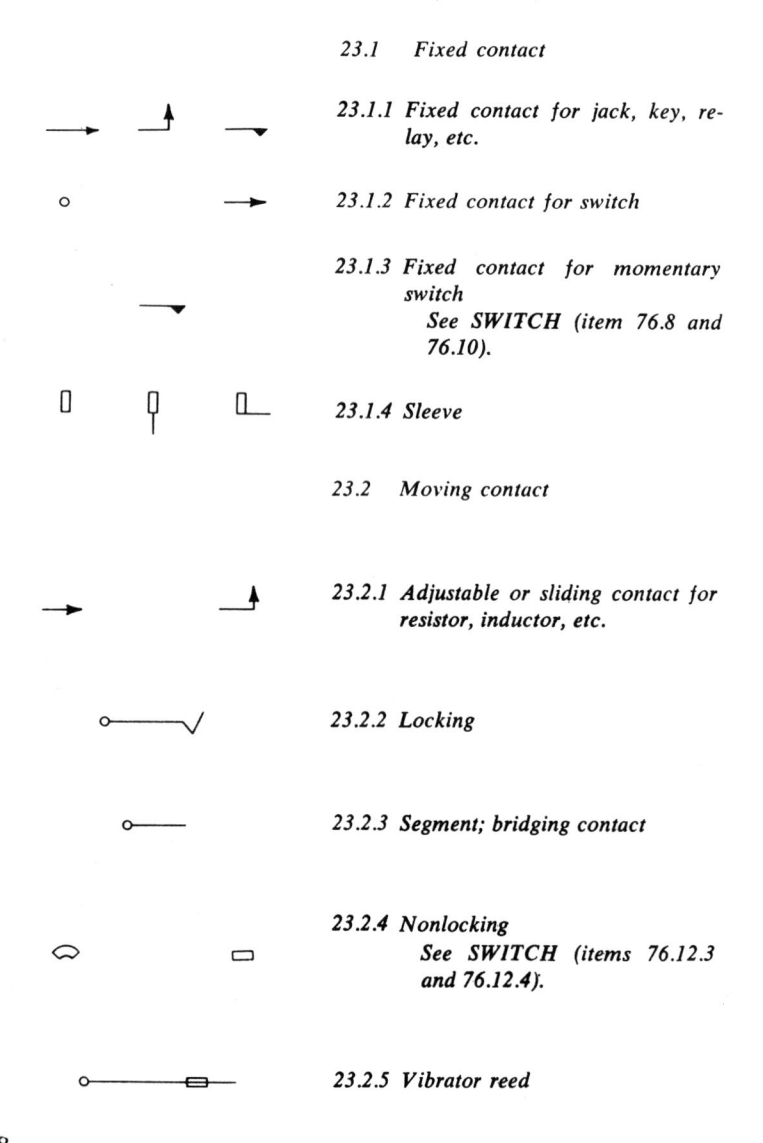

        23.1   Fixed contact

23.1.1 Fixed contact for jack, key, relay, etc.

23.1.2 Fixed contact for switch

23.1.3 Fixed contact for momentary switch
      See SWITCH (item 76.8 and 76.10).

23.1.4 Sleeve

        23.2   Moving contact

23.2.1 Adjustable or sliding contact for resistor, inductor, etc.

23.2.2 Locking

23.2.3 Segment; bridging contact

23.2.4 Nonlocking
      See SWITCH (items 76.12.3 and 76.12.4).

23.2.5 Vibrator reed

Electrical Symbols

23.2.6 Vibrator split reed

23.2.7 Rotating contact (slip ring) and brush

23.3 Basic contact assemblies

The standard method of showing a contact is by a symbol indicating the circuit condition it produces when the actuating device is in the deenergized or nonoperated position. The actuating device may be of a mechanical, electrical, or other nature, and a clarifying note may be necessary with the symbol to explain the proper point at which the contact functions, for example, the point where a contact closes or opens as a function of changing pressure, level, flow, voltage, current, etc. In cases where it is desirable to show contacts in the energized or operated condition and where confusion may result, a clarifying note shall be added to the drawing. Auxiliary switches or contacts for circuit breakers, etc., may be designated as follows:

(a) Closed when device is in energized or operated position,

(b) Closed when device is in de-energized or nonoperated position,

(aa) Closed when operating mechanism of main device is in energized or operated position,

(bb) Closed when operating mechanism of main device is in de-energized or nonoperated position.

See American Standard C37.2-1962 for further details.

In the parallel-line contact symbols showing the length of the parallel lines shall be approximately 1¼ times the width of the gap (except for item 23.6)

ELECTRICAL SYMBOLS

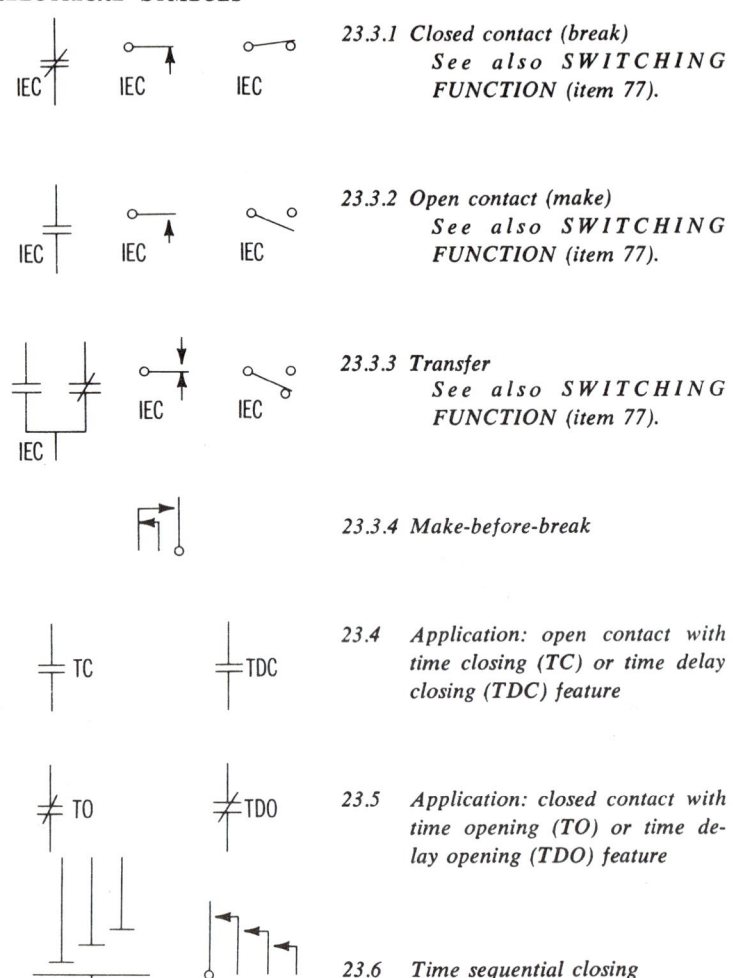

23.3.1 Closed contact (break)
See also SWITCHING FUNCTION (item 77).

23.3.2 Open contact (make)
See also SWITCHING FUNCTION (item 77).

23.3.3 Transfer
See also SWITCHING FUNCTION (item 77).

23.3.4 Make-before-break

23.4 Application: open contact with time closing (TC) or time delay closing (TDC) feature

23.5 Application: closed contact with time opening (TO) or time delay opening (TDO) feature

23.6 Time sequential closing

24 Contactor

See also RELAY (item 66).

Fundamental symbols for contacts, coils, mechanical connections, etc., are the basis of contactor symbols and should be used to represent contactors on complete diagrams. Complete diagrams of contactors consist of combinations of fundamental symbols

Electrical Symbols

for control coils, mechanical connections, etc., in such configurations as to represent the actual device.

Mechanical interlocking should be indicated by notes.

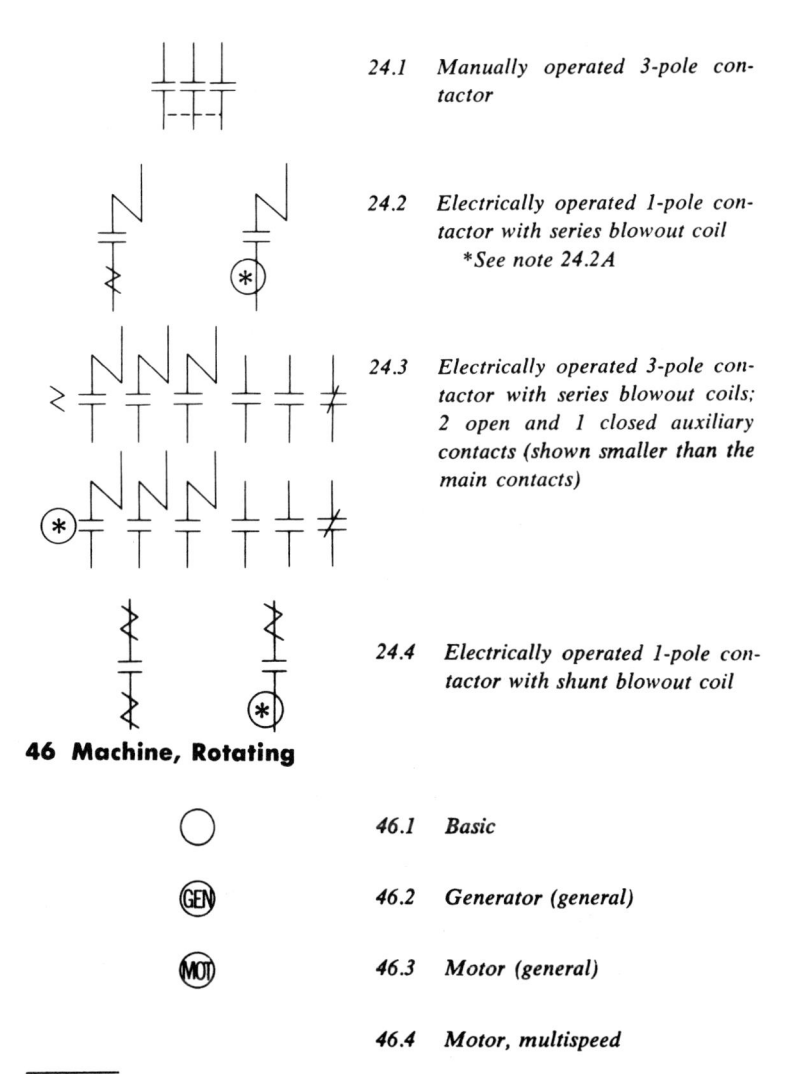

24.1 Manually operated 3-pole contactor

24.2 Electrically operated 1-pole contactor with series blowout coil
*See note 24.2A

24.3 Electrically operated 3-pole contactor with series blowout coils; 2 open and 1 closed auxiliary contacts (shown smaller than the main contacts)

24.4 Electrically operated 1-pole contactor with shunt blowout coil

46 Machine, Rotating

46.1 Basic

46.2 Generator (general)

46.3 Motor (general)

46.4 Motor, multispeed

Note 24.2A—The asterisk is not a part of the symbol. Always replace the asterisk by a device designation.

ELECTRICAL SYMBOLS

USE BASIC MOTOR SYMBOL AND NOTE SPEEDS

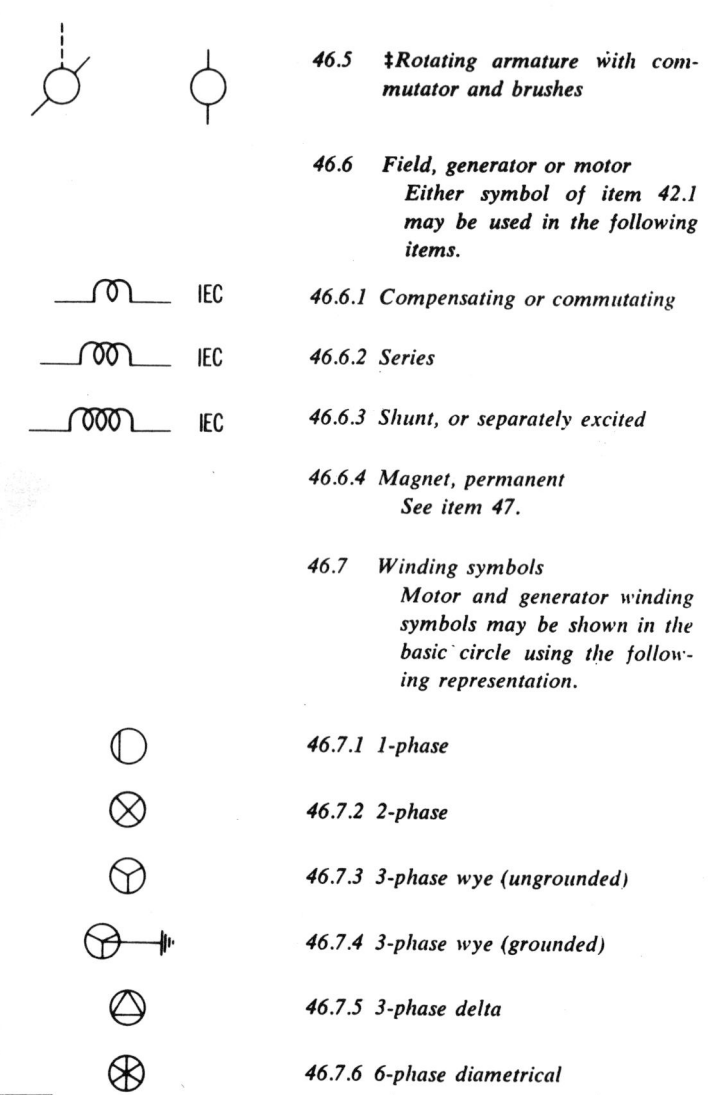

46.5 ‡*Rotating armature with commutator and brushes*

46.6 *Field, generator or motor*
Either symbol of item 42.1 may be used in the following items.

IEC   46.6.1 *Compensating or commutating*

IEC   46.6.2 *Series*

IEC   46.6.3 *Shunt, or separately excited*

46.6.4 *Magnet, permanent*
See item 47.

46.7 *Winding symbols*
Motor and generator winding symbols may be shown in the basic circle using the following representation.

46.7.1 *1-phase*

46.7.2 *2-phase*

46.7.3 *3-phase wye (ungrounded)*

46.7.4 *3-phase wye (grounded)*

46.7.5 *3-phase delta*

46.7.6 *6-phase diametrical*

‡The broken line - —— indicates where line connection to a symbol is made and is not a part of the symbol.

Electrical Symbols

46.7.7 6-phase double-delta

46.8 Direct-current machines; applications

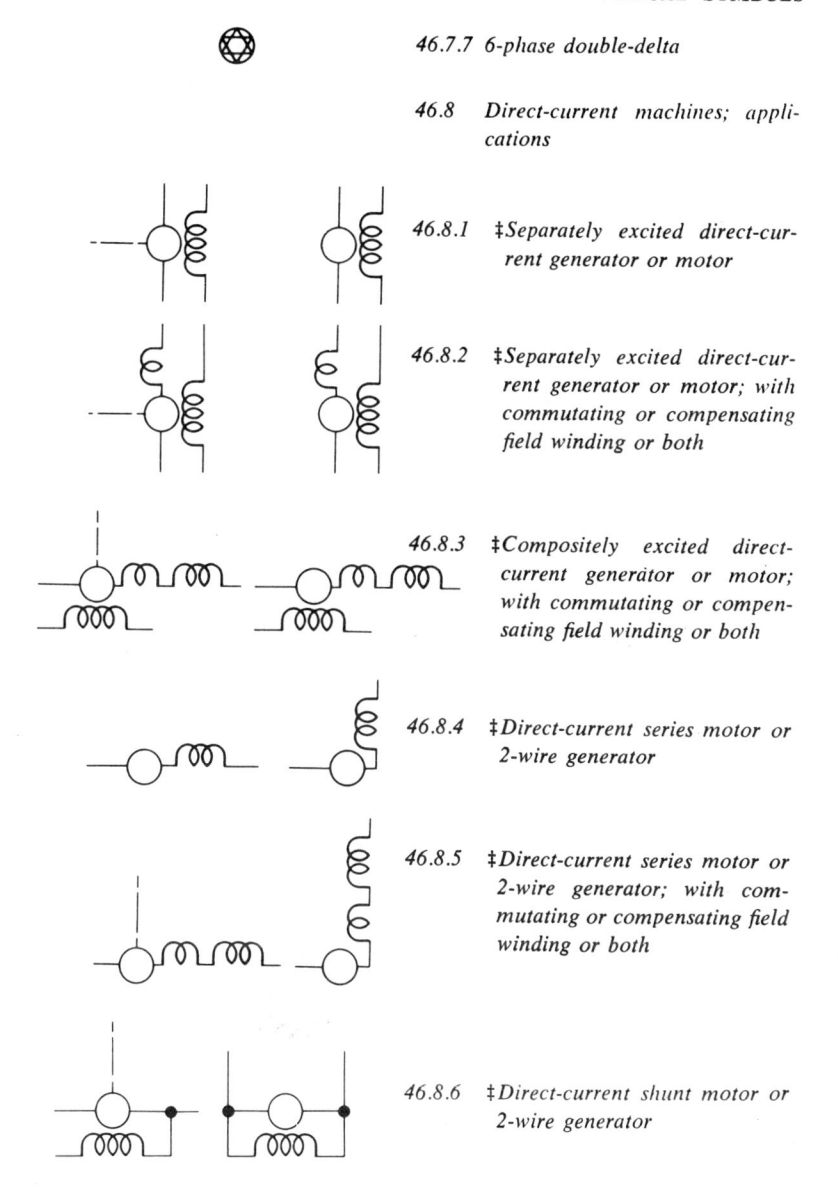

46.8.1 ‡Separately excited direct-current generator or motor

46.8.2 ‡Separately excited direct-current generator or motor; with commutating or compensating field winding or both

46.8.3 ‡Compositely excited direct-current generator or motor; with commutating or compensating field winding or both

46.8.4 ‡Direct-current series motor or 2-wire generator

46.8.5 ‡Direct-current series motor or 2-wire generator; with commutating or compensating field winding or both

46.8.6 ‡Direct-current shunt motor or 2-wire generator

---

‡The broken line - —— indicates where line connection to a symbol is made and is not a part of the symbol.

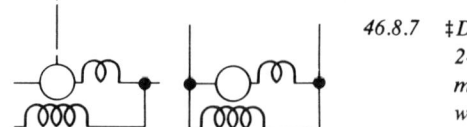

Electrical Symbols

46.8.7 ‡*Direct-current shunt motor or 2-wire generator; with commutating or compensating field winding or both*

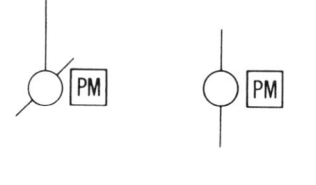

46.8.8 ‡*Direct-current permanent-magnet-field generator or motor*

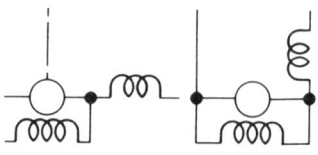

46.8.9 ‡*Direct-current compound motor or 2-wire generator or stabilized shunt motor*

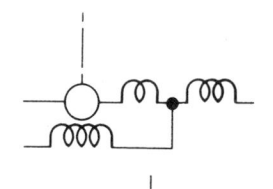

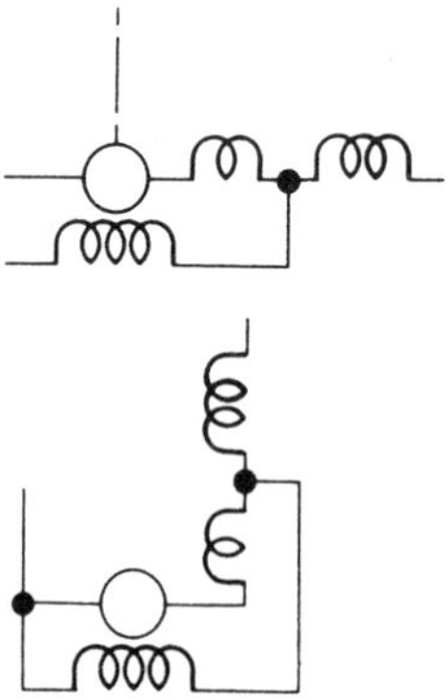

46.8.10 ‡*Direct-current compound motor or 2-wire generator or stabilized shunt motor; with commutating or compensating field winding or both*

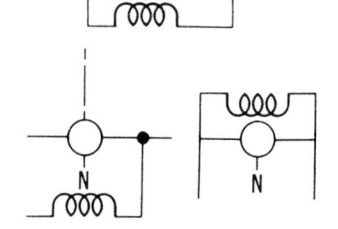

46.8.11 ‡*Direct-current 3-wire shunt generator*

‡The broken line - - - indicates where line connection to a symbol is made and is not a part of the symbol.

34

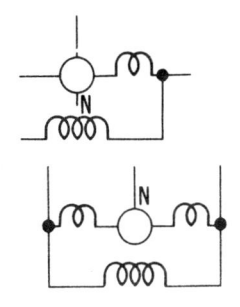

ELECTRICAL SYMBOLS

46.8.12 ‡*Direct-current 3-wire shunt generator; with commutating or compensating field winding or both*

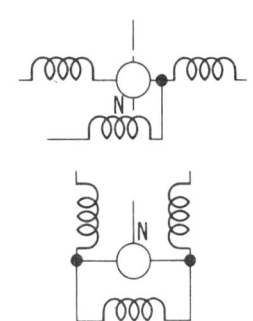

46.8.13 ‡*Direct-current 3-wire compound generator*

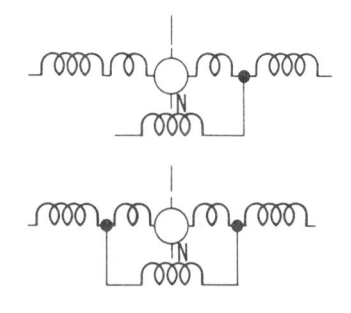

46.8.14 ‡*Direct-current 3-wire compound generator; with commutating or compensating field winding or both*

‡The broken line - —— indicates where line connection to a symbol is made and is not a part of the symbol.

35

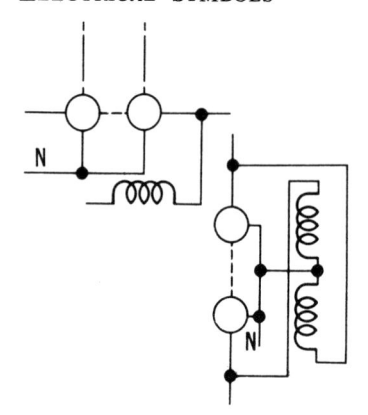

ELECTRICAL SYMBOLS

46.8.15 ‡*Direct-current balancer, shunt wound*

46.9 *Alternating-current machines; applications*

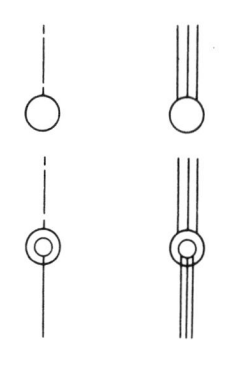

46.9.1 ‡*Squirrel-cage induction motor or generator, split-phase induction motor or generator, rotary phase converter, or repulsion motor*

46.9.2 ‡*Wound-rotor induction motor, synchronous induction motor, induction generator, or induction frequency converter*

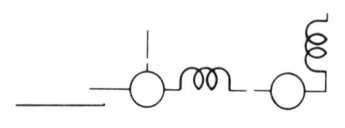

46.9.3 ‡*Alternating-current series motor*

48 Meter Instrument

Note 48A—*The asterisk is not a part of the symbol. Always replace the asterisk by one of the following letter combinations, depending on the function of the meter or instrument, unless some other identification is provided in the circle and explained on the diagram.*

‡The broken line - —indicates where line connection to a symbol is made and is not a part of the symbol.

ELECTRICAL SYMBOLS

A	*Ammeter IEC*
AH	*Ampere-hour*
CMA	*Contact-making (or breaking) ammeter*
CMC	*Contact-making (or breaking) Clock*
CMV	*Contact-making (or breaking) voltmeter*
CRO	*Oscilloscope or cathode-ray oscilograph*
DB	*DB (decibel) meter*
DBM	*DBM (decibels referred to 1 milliwatt) meter*
DM	*Demand meter*
DTR	*Demand-totalizing relay*
F	*Frequency meter*
G	*Galvanometer*
GD	*Ground detector*
I	*Indicating*
INT	*Integrating*
µA or	
UA	*Microammeter*
MA	*Milliammeter*
NM	*Noise meter*
OHM	*Ohmmeter*
OP	*Oil pressure*
OSCG	*Oscillograph string*
PH	*Phasemeter*
PI	*Position indicator*
PF	*Power factor*
RD	*Recording demand meter*
REC	*Recording*
RF	*Reaction factor*
SY	*Synchroscope*
TLM	*Telemeter*
T	*Temperature meter*
THC	*Thermal converter*

ELECTRICAL SYMBOLS
- *TT*   *Total time*
- *V*   *Voltmeter*
- *VA*   *Volt-ammeter*
- *VAR*   *Varmeter*
- *VARH*   *Varhour meter*
- *VI*   *Volume indicator; meter, audio level*
- *VU*   *Standard volume indicator; meter, audio lever*
- *W*   *Wattmeter*
- *WH*   *Watthour meter*

58 Path, Transmission, Conductor, Cable, Wiring

58.1 Guided path, general
A single line represents the entire group of conductors or the transmission path needed to guide the power or the signal. For coaxial and waveguide work, the recognition symbol is used at the beginning and end of each kind of transmission path and at intermediate points as needed for clarity. In waveguide work, mode may be indicated.

58.2 Conductive path or conductor; wire

58.2.1 Two conductors or conductive paths of wires

58.2.2 Three conductors or conductive paths of wires

58.2.3 "n" conductors or conductive paths of wires
            "n" conductors

58.5 Crossing of paths or conductors not connected
The crossing is not necessarily at a 90-degree angle.

Electrical Symbols

58.6   Junction of paths or conductors

• IEC

58.6.1   Junction (if desired)

———•——— IEC

58.6.1.1   Application: junction of paths, conductor, or cable. If desired indicate path type, or size

———Splice•——— IEC

58.6.1.2   Application: splice (if desired) of same size cables. Junction of conductors of same size or different size cables. If desired indicate sizes of conductors

⊢•—•⊣ IEC

58.6.2   Junction of connected paths, conductors, or wires

——┬—— IEC

OR ONLY IF REQUIRED BY SPACE LIMITATION

—•— IEC

63 Polarity Symbol

+ IEC   63.1 Positive

− IEC   63.2 Negative

76 Switch
See also FUSE (item 36); CONTACT, ELECTRIC (item 23); and DRAFTING PRACTICES (items 0.4.6 and 0.4.7).

Electrical Symbols

Fundamental symbols for contacts, mechanical connections, etc., may be used for switch symbols.

The standard method of showing switches is in a position with no operating force applied. For switches that may be in any one of two or more positions with no operating force applied and for switches actuated by some mechanical device (as in air-pressure, liquid-level, rate-of-flow, etc., switches), a clarifying note may be necessary to explain the point at which the switch functions.

When the basic switch symbols in items 76.1 through 76.4 are shown on a diagram in the closed position, terminals must be added for clarity.

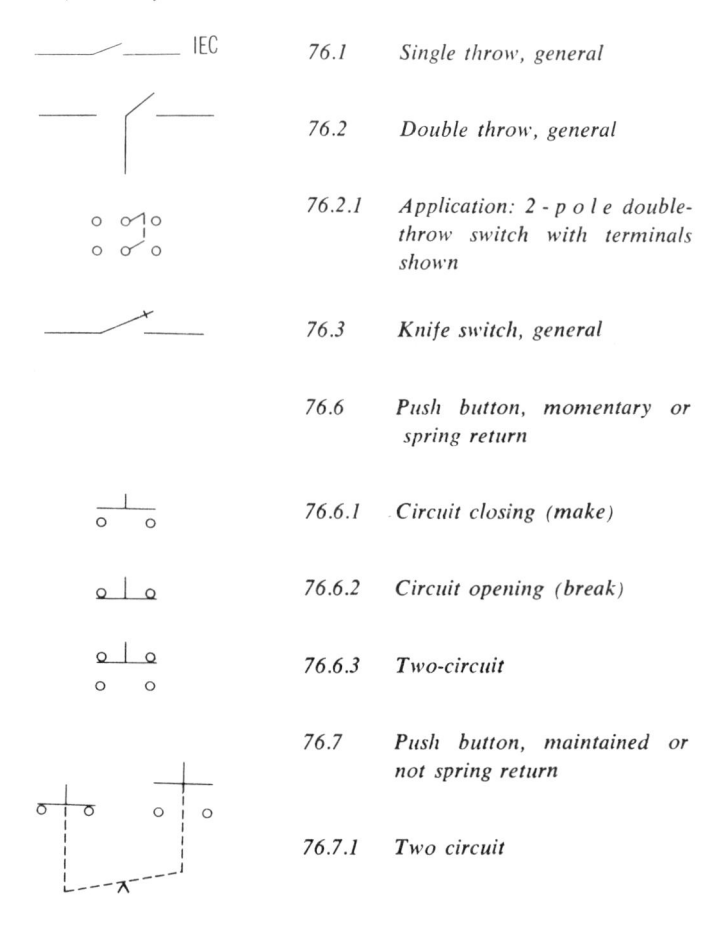

IEC	76.1	Single throw, general
	76.2	Double throw, general
	76.2.1	Application: 2-p o l e double-throw switch with terminals shown
	76.3	Knife switch, general
	76.6	Push button, momentary or spring return
	76.6.1	Circuit closing (make)
	76.6.2	Circuit opening (break)
	76.6.3	Two-circuit
	76.7	Push button, maintained or not spring return
	76.7.1	Two circuit

ELECTRICAL SYMBOLS

86 Transformer

86.1 General

IEC

Either winding symbol may be used. In the following items, the left symbol is used. Additional windings may be shown or indicated by a note. For power transformers use polarity marking H_1, X_1, etc., from American Standard C6.1-1956.

For polarity markings on current and potential transformers, see items 86.16.1 and 86.17.1.
In coaxial and waveguide circuits, this symbol will represent a taper or step transformer without mode change.

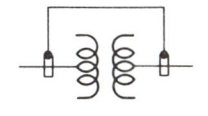

86.1.1 Application: transformer with direct-current connections and mode suppression between two rectangular waveguides

86.2 If it is desired especially to distinguish a magnetic-core transformer

86.2.1 Application: shielded transformer with magnetic core shown

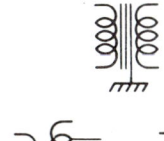

86.2.2 Application: transformer with magnetic core shown and with a shield between windings. The shield is shown connected to the frame

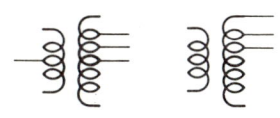

86.6 With taps, 1-phase

41

Electrical Symbols

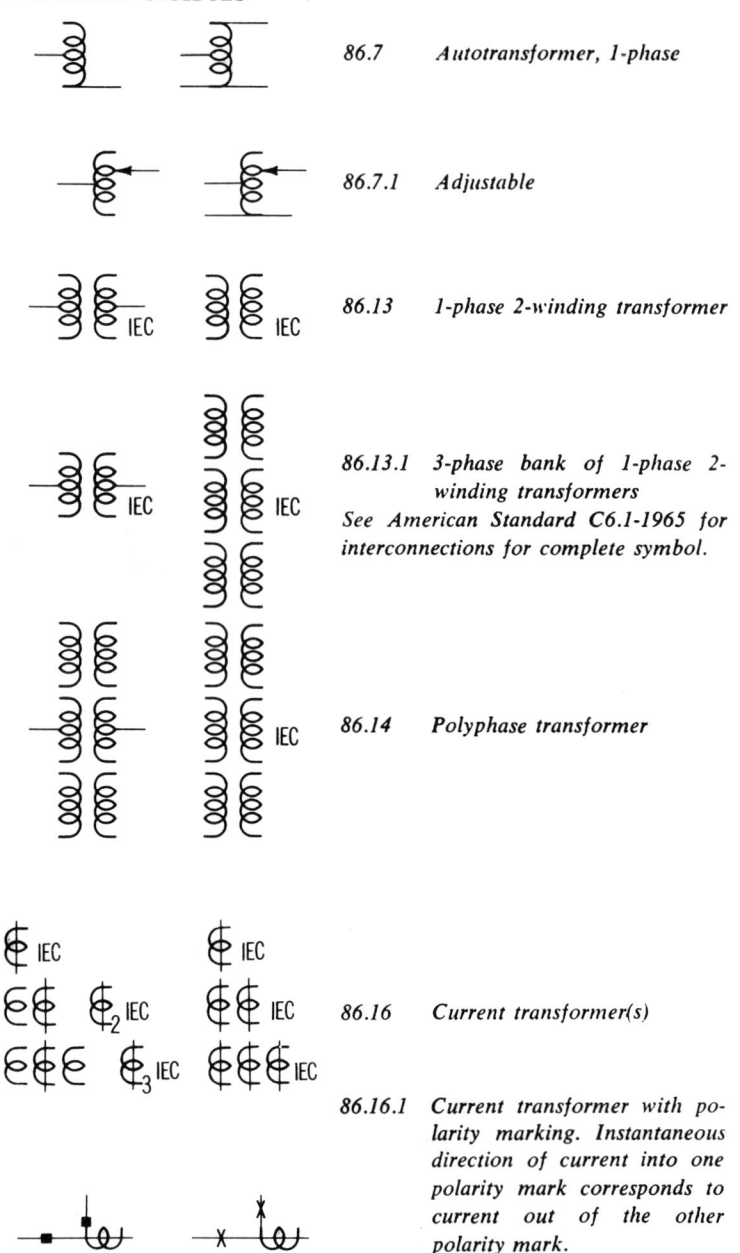

		86.7	Autotransformer, 1-phase
		86.7.1	Adjustable
		86.13	1-phase 2-winding transformer
		86.13.1	3-phase bank of 1-phase 2-winding transformers

See American Standard C6.1-1965 for interconnections for complete symbol.

		86.14	Polyphase transformer
		86.16	Current transformer(s)
		86.16.1	Current transformer with polarity marking. Instantaneous direction of current into one polarity mark corresponds to current out of the other polarity mark.

ELECTRICAL SYMBOLS

Symbol used shall not conflict with item 77 when used on same drawing.

86.16.2‡ Bushing-type current transformer

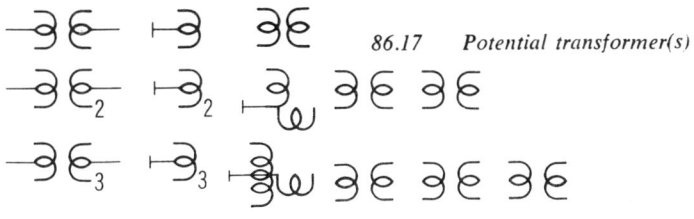

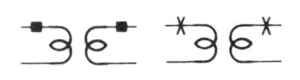

86.17 Potential transformer(s)

86.17.1 Potential transformer with polarity mark. Instantaneous direction of current into one polarity mark corresponds to current out of the other polarity mark.

Symbol used shall not conflict with item 77 when used on same drawing.

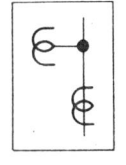

86.18 Outdoor metering device

86.19 Transformer winding connection symbols
For use adjacent to the symbols for the transformer windings.

IEC

86.19.1 2-phase 3-wire, ungrounded

IEC

86.19.1.1 2-phase 3-wire, grounded

43

Electrical Symbols

Symbol	Reference	Description
─┼─ IEC	86.19.2	2-phase 4-wire
─┼─ (grounded) IEC	86.19.2.1	2-phase 5-wire, grounded
△ IEC	86.19.3	3-phase 3-wire, delta or mesh
△ (grounded) IEC	86.19.3.1	3-phase 3-wire, delta, grounded
△ with stem	86.19.4	3-phase 4-wire, *delta*, ungrounded
△ with stem grounded	86.19.4.1	3-phase 4-wire, delta, grounded
∠ IEC	86.19.5	3-phase, open-delta
∠ (grounded) IEC	86.19.5.1	3-phase, open-delta, grounded at common point
∠ (mid grounded)	86.19.5.2	3-phase, open-delta, grounded at middle point of one transformer
⊿ IEC	86.19.6	3-phase, broken-delta
⋋ IEC	86.19.7	3-phase, wye or star, ungrounded
⋋ (grounded) IEC	86.19.7.1	3-phase, wye, grounded neutral. The direction of the stroke representing the neutral can be arbitrarily chosen.
⋏ IEC	86.19.8	3-phase 4-wire, ungrounded

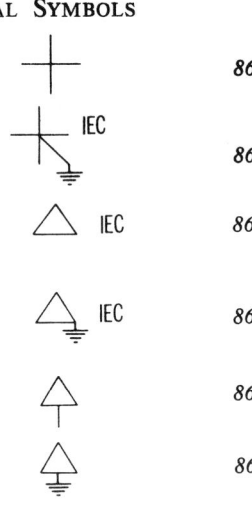

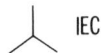

CHAPTER 1

Review Definitions

Definitions are covered in Article 100 of the NEC (National Electrical Code). The questions that follow will not cover all of the definitions but only the more pertinent ones.

1-1 What is "accessible" (as applied to wiring methods)?
Readily available to inspection, repair, removal, etc., without disturbing the building structure or finish. Not permanently closed by structure or finish of buildings.

1-2 What is "accessible" (as applied to equipment)?
May be readily reached without climbing over obstacles, not in locked or other hard-to-get-at areas, e.g., panels in kitchen cupboards, which are mounted in or on the walls above washers and dryers or in closets or bathrooms. Service entrance equipment that can only be reached by going into a basement, behind a stairway, or some other hindrance would most certainly not be termed "accessible."

1-3 What is a "fixed" appliance?
One that is fastened in a definite location, e.g., a garbage disposal, built-in dishwasher, built-in electric heater, built-in cooking equipment, etc.

1-4 What is a "portable" appliance?
One that is definitely easily moved from one location to another, e.g., mixers, hair dryers, portable dishwashers, portable electric heaters, radios, etc.

1-5 What is a "stationary" appliance?
One which in normal use is not easily moved, e.g., automatic clothes washer, window air conditioner, etc.

REVIEW DEFINITIONS

1-6 What does "approved" mean?

Any appliance, wiring material, or other electrical equipment that is acceptable to the enforcing authorities. Frequently the Underwriters Laboratory is the most acceptable authority to inspectors. Do not be fooled by a UL label on the motor or cord of an appliance without such a label on the entire article.

1-6a What does "Approved for the Purpose" mean?

Approved for some specific application, purpose, environment, described in a particular Code requirement.

1-7 What is a branch circuit?

A branch circuit is the portion of a wiring system that extends beyond the last overcurrent-protection device. In interpreting this, you must not consider the thermal cutout or the motor overload protection as the beginning of the branch circuit. The branch circuit actually begins at the final fusing or circuit-breaker point where the circuit breaks off to supply the motor.

1-8 What is an appliance branch circuit?

This is the circuit supplying one or more outlets connecting appliances only; there is no permanently connected lighting on this circuit, except the lighting that may be built into the appliance. This term is most often used in connection with Article 220-3b of the NEC, which refers to outlets for small appliance loads in kitchens, laundries, pantries, and dining and breakfast rooms of dwellings.

1-9 What is a general-purpose outlet?

This is a branch circuit to which lighting and/or appliances may be connected. Lighting may be connected to this circuit, whereas in 1-8, lighting cannot be connected.

1-10 What is a multiwire branch circuit?

A multiwire branch circuit is one that has two or more ungrounded conductors with a potential difference between them and also has an identified grounding conductor with an equal potential difference between it and each of the other wires; e.g., a three-wire 120/240-volt system or a 120/208-volt Wye system, using two- or three-phase conductors and a grounded conductor. However, in either case, the "hot" wires must not be tied to one phase but must be connected to different phases to make the system a multiwire circuit.

REVIEW DEFINITIONS

1-11 What is a circuit breaker?

A device that is designed not only to open and close a circuit nonautomatically but also to open the circuit automatically at a predetermined current-overload value. The circuit breaker may be thermally or magnetically operated; however, ambient temperatures affect the operation of the thermally operated type, so that the trip value of the current is not as stable as with the magnetic type.

1-12 What is a current-carrying conductor?

A conductor that normally carries current.

1-13 What is a noncurrent-carrying conductor?

One that carries current only in the event of a malfunction of equipment or wiring. An equipment ground is a good example; it is employed for protection and is quite a necessary part of the wiring system, but it is not used for carrying normal currents, except in the case of faulty operation where it aids in tripping the overload-protection device.

1-14 What is a pressure connector (solderless)?

A device that establishes a good electrical connection between two or more conductors by some means of mechanical pressure. It is used in place of soldering connections and should be of the approved type.

1-15 What is meant by "demand factor"?

This is the ratio between the maximum demand on a system or part of a system and the total connected load on the same system or part of the system.

1-16 What is meant by "dust-tight"?

Being capable of keeping dust out of the enclosing case so that dust cannot interfere with normal operation. This is discussed further in connection with Articles 500 and 502 of the NEC, which both cover hazardous locations.

1-17 What is meant by "explosion-proof application"?

Apparatus enclosed in a case that is capable of sustaining an explosion that may occur within itself and is also capable of preventing ignition of specified gases or vapors surrounding the enclosure by sparks, flashes, or explosion of the gases or vapors within; it must also operate at a temperature that will

47

REVIEW DEFINITIONS

not ignite any inflammable atmosphere or residue surrounding it. If an explosion does occur within the equipment, the gases are allowed to escape either by a ground joint or by threads, and the escaping gases are thereby cooled to a temperature low enough to inhibit the ignition of any external gases.

1-18 What is a feeder?

The circuit conductors between the service-entrance equipment or isolated generating plant and the branch circuit overload device or devices. Generally, feeders are comparatively large in size and supply a feeder panel, which is composed of a number of branch-circuit overload devices.

1-19 What is a fitting?

A mechanical device, such as a locknut or bushing, that is intended primarily for a mechanical rather than an electrical function.

1-20 What is meant by a "ground"?

An electrical connection, either accidental or intentional, that exists between an electrical circuit or equipment and earth, or some other electrical conducting body which serves in place of the earth.

1-21 What does "grounded" mean?

Connected to earth or to some other conducting body which serves in place of the earth.

1-22 What is a grounded conductor?

A system or circuit conductor which is intentionally grounded.

1-23 What is a grounding conductor?

A conductor that is used to connect equipment, devices, or wiring systems with grounding electrodes.

1-23a What is a grounding conductor, equipment?

It is the conductor used to connect noncurrent-carrying metal parts of equipment, raceways and other enclosures to the system grounding conductor at the service and/or the grounding electrode conductor.

1-23b What is a grounding electrode conductor?

It is a conductor used to connect the grounding electrode to the equipment grounding conductor and/or to the grounded conductor of the circuit at the service.

Review Definitions

1-24 What does "identified" mean?

This means that a conductor or terminal is so marked that it is recognizable as being grounded. The neutral is referred to as the identified conductor.

1-25 What is an outlet?

A point in the wiring system at which current is taken to be used in some equipment.

1-26 What is meant by "rain-tight"?

Capable of withstanding a beating rain without resulting in the entrance of water.

1-26a What is a receptacle?

A receptacle is a contact device installed at the outlet for the connection of a single attachment plug.

A single receptacle is a single device with no other contact device on the same yoke. A multiple receptacle is a single device containing two or more receptacles.

1-26b What is a rainproof?

So constructed, protected or treated as to prevent rain from interfering with successful operation of the apparatus.

Pay particular attention to the following questions; they involve services and are probably among the most misused of any definitions in the NEC.

1-27 What is meant by the term "service"?

The conductors and equipment for delivering electrical energy from the secondary distribution system—the street main, the distribution feeder, or the transformer—to the wiring system on the premises. This includes the service-entrance equipment and the grounding electrode.

1-28 What are service conductors?

The portion of the supply conductors that extend from the street main, duct, or transformers to the service-entrance equipment of the premises supplied. For overhead conductors, this includes the conductors from the last line pole (this does not mean the service pole) to the service equipment.

1-29 What is service cable?

Service conductors that are made up in the form of a cable and are normally referred to as ASE cable, SE cable, or USE

REVIEW DEFINITIONS

cable. A new cable, called Tri-Plex, has been recently added; it consists of two insulated wires and one bare wire twisted together without an outer covering over the three wires.

1-30 What is meant by "service drop"?

The overhead conductors from the last pole or other aerial support to and including the splices, if any, connecting to the service entrance conductors at the building or other structure. If there is a service pole with a meter on it, such as a farm service pole, the service drop does not stop at the service pole; all wires extending from this pole to a building or buildings are service drops, as well as the conductors from the last line pole to the service pole. (See NEC, Article 100.)

1-31 What are service-entrance conductors (overhead system)?

That portion of the service conductors between the terminals of service equipment and a point outside the building, clear of buildings walls, where they are joined by a splice or tap to the service drop, street main, or other source of supply.

1-32 What are service-entrance conductors (underground system)?

The service conductors between the terminals of the service equipment and the point of connection to the service lateral. Where service equipment is located outside the building walls, there may be no service-entrance conductors, or they may be entirely outside the building.

1-32a What are "Service-Entrance Conductors Sub-Sets?

Sub-sets of service-entrance conductors are taps from main service conductors, run to service-equipment.

1-33 What is meant by "service equipment"?

The necessary equipment, usually consisting of circuit breakers or switches and fuses, and their accessories, located near the point of entrance of supply conductors to a building or other structure, or an otherwise defined area, and intended to constitute the main control and means of cutoff of the supply.

1-34 What is meant by "service lateral"?

The underground service conductors between the street main, including any risers at the pole or other structure or from transformers, and the first point of connection to the service-

Review Definitions

entrance conductors in a terminal box. The point of connection is considered to be the point of entrance of the service conductors into the building.

1-35 What is meant by "service raceway"?
The rigid metal conduit, electrical metallic tubing, or other raceway that encloses service-entrance conductors.

1-36 What is meant by "special permission"?
The written consent of the authority enforcing the NEC.

1-37 What is meant by a "general-use switch"?
A device intended for use as a switch in general distribution and branch circuits. It is rated in amperes and is capable of interrupting its rated current at its rated voltage.

1-38 What is meant by a "T-rated switch"?
An AC general-use snap switch that is capable of use on resistive and inductive loads that do not exceed the ampere rating at the voltage involved, on tungsten-filament lighting loads that do not exceed the ampere rating at 120 volts, and on motor loads that do not exceed 80% of their ampere rating at the rated voltage.

1-39 What is meant by an "isolating switch"?
A switch that is intended for isolating an electric circuit from its source of power. It has no interrupting rating and is intended to be operated only after the circuit has been opened by some other means.

1-40 What is meant by a "motor-circuit switch"?
A switch, rated in horsepower, that is capable of interrupting the maximum operating overload current of a motor of the same horsepower rating as the switch at the rated voltage.

1-41 What is meant by "water-tight"?
So constructed that moisture will not enter the enclosing case.

1-42 What is meant by "weatherproof"?
So constructed or protected that exposure to the weather will not interfere with successful operation. Rain-tight or water-tight may fulfill the requirements for "weatherproof." However, weather conditions vary, and consideration should be given to the conditions resulting from snow, ice, dust, and temperature extremes.

REVIEW DEFINITIONS

1-43 What is meant by the "voltage" of a circuit?

This is the greatest effective difference of potential (root-mean-square difference of potential) that exists between any two conductors of a circuit. On various systems, such as 3-phase 4-wire, single-phase 3-wire, and 3-wire direct current, there may be various circuits of numerous voltages.

There are a number of other definitions given under Article 100 of the NEC. It would be well to become thoroughly familiar with all of them. Do not hesitate to refer to them from time to time.

CHAPTER 2

Ohm's Law and Other Electrical Formulas

When a current flows in an electric circuit, the magnitude of the current is determined by dividing the electromotive force (volts, designated by the letter E) in the circuit by the resistance (ohms, designated by the letter R) of the circuit; the resistance is depenent on the material, cross section, and length of the conductor. The current is measured in amperes and is designated by the letter I. The relationship between an unvarying electric current (I), the electromotive force (E), and the resistance (R) is expressed by Ohm's law. The following equations take into consideration only resistance; therefore, they are customarily known as the DC formulas for Ohm's law. However, in most calculations for AC circuits, which is the ordinary wiring application, this formula is quite practical to use. In a later portion of this book, another form of Ohm's law, which deals with inductive and capacitive reactance, will be discussed.

2-1 State the three equations for Ohm's law and explain what the letters in the formulas mean.

$$I = \frac{E}{R} \qquad R = \frac{E}{I} \qquad E = IR$$

where,
   I is the current flow in amperes,
   E is the electromotive force in volts,
   R is the resistance in ohms.

2-2 A direct-current circuit has a resistance of 5 ohms. If

OHM'S LAW AND OTHER ELECTRICAL FORMULAS

a voltmeter connected across the terminals of the circuit reads 10 volts, how much current is flowing?

From Ohm's law, the current is

$$I = \frac{E}{R} = \frac{10}{5} = 2 \ amps$$

2-3 If the resistance of a circuit is 25 ohms, what voltage is necessary for a current flow of 4 amperes?

From Ohm's law

$$E = I \times R = 4 \times 25 = 100 \ volts$$

2-4 If the potential across a circuit is 40 volts and the current is 5 amperes, what is the resistance?

From Ohm's law

$$R = \frac{E}{I} = \frac{40}{5} = 8 \ ohms$$

SERIES CIRCUITS

A series circuit may be defined as one in which the resistive elements are connected in a continuous run (i.e., connected end to end) as shown in Fig. 1. It is evident that since the circuit has no branches, the same current flows in each resistance. The total

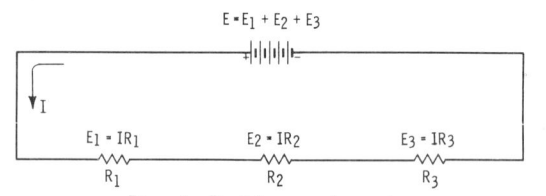

Fig. 1. Resistances in series.

potential across the entire circuit equals the sum of potential drops across each individual resistance, or

$$E_1 = IR_1$$
$$E_2 = IR_2$$
$$E_3 = IR_3$$
$$E = E_1 + E_2 + E_3$$

Ohm's Law and Other Electrical Formulas

and
$$R = R_1 + R_2 + R_3$$

The equation for the total potential of the circuit is

$$E = IR_1 + IR_2 + IR_3 = I(R_1 + R_2 + R_3)$$

and

$$I = \frac{E}{R_1 + R_2 + R_3} = \frac{E}{R}$$

2-5 If the individual resistances shown in Fig. 1 are 5, 10, and 15 ohms, respectively, what potential must the battery supply to force a current of 0.5 ampere through the circuit?

The total resistance is
$$R = 5 + 10 + 15 = 30 \; ohms$$

Hence, the total voltage is
$$E = 0.5 \times 30 = 15 \; volts$$

As a check, we can calculate the individual voltage drop across each part.
$$E_1 = 0.5 \times \;\; 5 = 2.5 \; volts$$
$$E_2 = 0.5 \times 10 = 5.0 \; volts$$
$$E_3 = 0.5 \times 15 = 7.5 \; volts$$

and
$$E = E_1 + E_2 + E_3 = 2.5 + 5.0 + 7.5 = 15 \; volts$$

2-6 In order to determine the voltage of a DC source, three resistance units of 10, 15, and 30 ohms are connected in series with this source. If the current through the circuit is 2 amperes, what is the potential of the source?

$$E = I(R_1 + R_2 + R_3) = 2(10 + 15 + 30)$$
$$E = 2 \times 55 = 110 \; volts$$

Ohm's Law and Other Electrical Formulas
PARALLEL CIRCUITS

In a parallel, or divided, circuit such as that shown in Fig. 2, the same voltage appears across each resistance in the group; the current flowing through each resistance is inversely proportional to the value of the resistance. The sum of all the currents, however, is equal to the total current leaving the battery. Thus:

$$E = I_1 R_1 = I_2 R_2 = I_3 R_3$$

and

$$I = I_1 + I_2 + I_3$$

When Ohm's law is applied to the individual resistances, the following equations are obtained:

$$I_1 = \frac{E}{R_1} \qquad I_2 = \frac{E}{R_2} \qquad I_3 = \frac{E}{R_3}$$

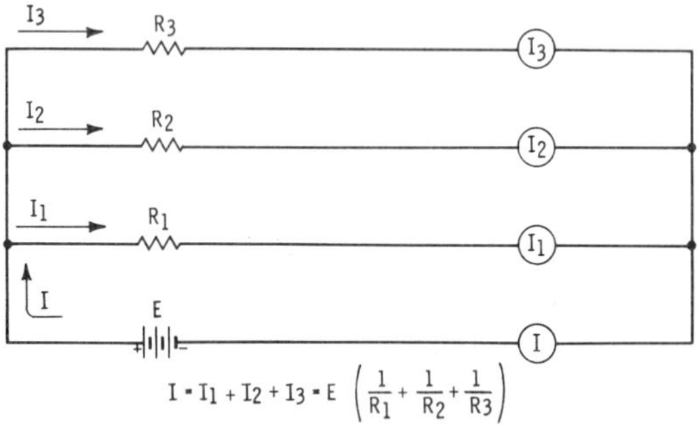

Fig. 2. Resistances in parallel.

Hence

$$I = \frac{E}{R_1} + \frac{E}{R_2} + \frac{E}{R_3}$$

or

$$I = E \left(\frac{1}{R_1} + \frac{1}{R_2} + \frac{1}{R_3} \right)$$

and, since $I = E/R$, the equivalent resistance of the several resistances connected in parallel is

Ohm's Law and Other Electrical Formulas

$$\frac{1}{R} = \frac{1}{R_1} + \frac{1}{R_2} + \frac{1}{R_3}$$

We have found, then, that any number of resistances in parallel can be replaced by an equivalent resistance whose value is equal to the reciprocal of the sum of the reciprocals of the individual resistances. You will find that the sum of resistances in parallel will always be smaller than the value of the smallest resistor in the group.

The value $1/R$, or the reciprocal of the value of the resistance, is expressed as the conductance of the circuit; its unit is mho, or ohm spelled backwards, and is usually expressed by g or G.

Where there are only two resistances connected in parallel

$$R = \frac{R_1 \times R_2}{R_1 + R_2} \; ohms$$

Where there are any number of equal resistances connected in parallel, you may divide the value of one resistance by the number of resistances.

2-7 **A resistance of 2 ohms is connected in series with a group of three resistances in parallel, which are 4, 5, and 20 ohms, respectively. What is the equivalent resistance of the circuit?**

The equivalent resistance of the parallel network is

$$\frac{1}{R} = \frac{1}{4} + \frac{1}{5} + \frac{1}{20} = \frac{5}{20} + \frac{4}{20} + \frac{1}{20} = \frac{10}{20} \; or \; 0.5 \; mho$$

$$R = \frac{1}{0.5} = 2 \; ohms$$

The circuit is now reduced to two series resistors of 2 ohms each, as shown in Fig. 3; the equivalent resistance of the circuit is 2 + 2, or 4 ohms.

2-8 **Two parallel resistors of 2 and 6 ohms are connected in series with a group of three parallel resistors of 1, 3, and 6 ohms, respectively. If the two parallel-resistance groups are connected in series by means of a 1.5-ohm resistor, what is the equivalent resistance of the system?**

OHM'S LAW AND OTHER ELECTRICAL FORMULAS

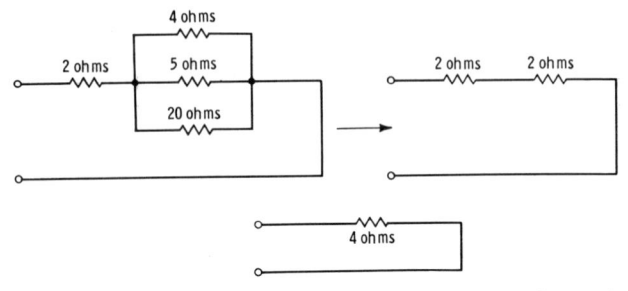

Fig. 3. *Equivalent resistance of the series-parallel circuit of question 2-7.*

Replace the 2- and 6-ohm resistors by a resistance of R_1, where

$$R_1 = \frac{2 \times 6}{2 + 6} = \frac{12}{8} = 1.5 \; ohms$$

Replace the group of three resistors by R_2, where

$$\frac{1}{R_2} = \frac{1}{1} + \frac{1}{3} + \frac{1}{6} = \frac{6}{6} + \frac{2}{6} + \frac{1}{6} = \frac{9}{6}$$

$$R_2 = \frac{6}{9} = 0.667 \; ohm$$

The circuit is now reduced to three series resistances, the values of which are 1.5, 1.5 and 0.667 ohms, as shown in Fig. 4. Their combined values are:

$$R_T = R + R_1 + R_2$$
$$R_T = 1.5 + 1.5 + 0.667 = 3.667 \; ohms$$

Fig. 4. *Equivalent resistance of the series-parallel circuit of question 2-8.*

Ohm's Law and Other Electrical Formulas
UNITS OF AREA AND RESISTANCE

The circular mil is the unit of cross section used in the American wire gauge (AWG) or the B&S wire gauge (Table 1). The term *mil* means one thousandth of an inch (0.001 inch). A circular mil is the area of a circular wire with a diameter of one mil, 0.001".

Table 1. Properties of Conductors

Size AWG MCM	Area Cir. Mils	Concentric Lay Stranded Conductors No. Wires	Concentric Lay Stranded Conductors Diam. Each Wire Inches	Bare Conductors Diam. Inches	Bare Conductors *Area Sq. Inches	D.C. Resistance Ohms/M FT. At 25°C. 77°F. Copper Bare Cond.	D.C. Resistance Ohms/M FT. At 25°C. 77°F. Copper Tin'd. Cond.	Aluminum
18	1620	Solid	.0403	.0403	.0013	6.51	6.79	10.7
16	2580	Solid	.0508	.0508	.0020	4.10	4.26	6.72
14	4110	Solid	.0641	.0641	.0032	2.57	2.68	4.22
12	6530	Solid	.0808	.0808	.0051	1.62	1.68	2.66
10	10380	Solid	.1019	.1019	.0081	1.018	1.06	1.67
8	16510	Solid	.1285	.1285	.0130	.6404	.659	1.05
6	26240	7	.0612	.184	.027	.410	.427	.674
4	41740	7	.0772	.232	.042	.259	.269	.424
3	52620	7	.0867	.260	.053	.205	.213	.336
2	66360	7	.0974	.292	.067	.162	.169	.266
1	83690	19	.0664	.332	.087	.129	.134	.211
0	105600	19	.0745	.372	.109	.102	.106	.168
00	133100	19	.0837	.418	.137	.0811	.0843	.133
000	167800	19	.0940	.470	.173	.0642	.0668	.105
0000	211600	19	.1055	.528	.219	.0509	.0525	.0836
250	250000	37	.0822	.575	.260	.0431	.0449	.0708
300	300000	37	.0900	.630	.312	.0360	.0374	.0590
350	350000	37	.0973	.681	.364	.0308	.0320	.0505
400	400000	37	.1040	.728	.416	.0270	.0278	.0442
500	500000	37	.1162	.813	.519	.0216	.0222	.0354
600	600000	61	.0992	.893	.626	.0180	.0187	.0295
700	700000	61	.1071	.964	.730	.0154	.0159	.0253
750	750000	61	.1109	.998	.782	.0144	.0148	.0236
800	800000	61	.1145	1.030	.833	.0135	.0139	.0221
900	900000	61	.1215	1.090	.933	.0120	.0123	.0197
1000	1000000	61	.1280	1.150	1.039	.0108	.0111	.0177
1250	1250000	91	.1172	1.289	1.305	.00863	.00888	.0142
1500	1500000	91	.1284	1.410	1.561	.00719	.00740	.0118
1750	1750000	127	.1174	1.526	1.829	.00616	.00634	.0101
2000	2000000	127	.1255	1.630	2.087	.00539	.00555	.00885

Ohm's Law and Other Electrical Formulas

* Area given is that of a circle having a diameter equal to the over-all diameter of a stranded conductor.

The values given in the Table are those given in Handbook 100 of the National Bureau of Standards except that those shown in the 8th column are those given in Specification B33 of the American Society for Testing and Materials, and those shown in the 9th column are those given in Standard No. S-19-81 of the Insulated Power Cable Engineers Association and Standard No. WC3-1964 of the National Electrical Manufacturers Association.

The resistance values given in the last three columns are applicable only to direct current. When conductors larger than No. 4/0 are used with alternating current the multiplying factors in Table 9, Chapter 9 should be used to compensate for skin effect.

The circular mil area of any solid cylindrical wire is equal to its diameter (expressed in mils) squared. For example, the area in circular mils (written CM or cir. mils) of a wire having a diameter of ⅜ inch (0.375) equals 375 × 375 = 375² = 140,625 CM. The diameter in mils of a solid circular wire is equal to the square root of its circular mil area. Assuming that a conductor

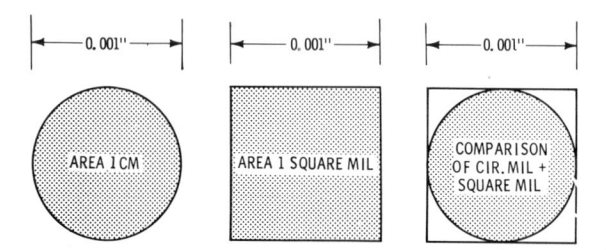

Fig. 5. Enlarged view of one circular mil and one square mil, with a comparison of the two.

has an area of 500,000 CM, its diameter in mils is the square root of 500,000, which is equal to 707 mils, or 0.707 inch (approximately). The area in square inches of a wire whose diameter is one mil is

$$\frac{\pi D^2}{4} = 0.7854 \times 0.001^2 = 0.0000007854 \text{ sq. in.}$$

The square mil is the area of a square whose sides are each one mil (0.001 inch). Hence, the area of a square mil is 0.001²,

OHM'S LAW AND OTHER ELECTRICAL FORMULAS

or 0.000001 square inch. With reference to the previous definitions of the circular mil and the square mil, it is obvious that in order to convert a unit of circular area into its equivalent area in square mils, the circular mil must be multiplied by $\pi/4$, or 0.7854, which is the same as dividing by 1.273. Conversely, to convert a unit area of square mils into its equivalent area in circular mils, the square mil should be divided by $\pi/4$, or 0.7854, which is the same as multiplying by 1.273.

The above relations may be written as follows:

$$square\ mils = circular\ mils \times 0.7854 = \frac{circular\ mils}{1.273}$$

$$circular\ mils = \frac{square\ mils}{0.7854} = square\ mils \times 1.273$$

Thus, any circular conductor may be easily converted into a rectangular conductor (a bus bar, for example) containing an equivalent area or current-carrying capacity.

2-9 If a No. 10 wire (B&S or AWG) has a diameter of 101.9 mils, what is its circular mil area?

$$area = 101.9^2 = 10{,}383.6\ CM$$

Referring to Table 1, you will find this to be approximately true. If you concentrate on remembering the CM area of No. 10 wire as 10,380 CM, you will find this invaluable in arriving at the CM area of any size wire without the use of a table. For practical purposes, if you should not have a wire table, such as Table 1, readily available, you may find the circular mil area of any wire size by juggling back and forth, and the answer will be close enough for all practical purposes.

Every three sizes removed from No. 10 doubles or halves in CM.

Every ten sizes removed from No. 10 is 1/10 or 10 times in CM.

Example:

Ten sizes larger than No. 10 is No. 0 wire; No. 10 wire has a circular mil area of 10,380 CM; therefore No. 0 would be 103,800 CM. Table 1 shows an area of 105,600 CM for No. 0 wire, or about a 1.7% error.

OHM'S LAW AND OTHER ELECTRICAL FORMULAS

Example:
Ten sizes smaller than No. 10 is No. 20 wire. Number 20 wire should then have a circular mil area of 1,038.1 CM.

You can see from the examples above that for quick figuring, the percentage of error is quite small; however, you should always have a wire table at your finger tips.

2-10 Number 000 wire (AWG) has an area of 167,800 CM. What is its diameter?

The diameter is

$$\sqrt{167,800} = 409.6 \text{ mils, or } 0.4096 \text{ inch}$$

2-11 A certain switchboard arrangement necessitates a conversion from a circular conductor to a rectangular bus bar having an equivalent area. If the diameter of the solid conductor measures 0.846 inch, calculate (a) the width of an equivalent bus bar, if the thickness of the bus bar is $\frac{1}{4}$ inch; (b) its area in circular mils; (c) the current-carrying capacity, if one square inch of copper carries 1,000 amperes.

The area, in square inches, of the circular conductor is

$$A = 0.7854 \times D^2 = 0.7854 \times 0.846^2 = 0.562 \text{ square inch}$$

The area, in square inches, of the bus bar is

$$A' = 0.25 \times W$$

where W is the width of the bus bar.
Since

$$A' = A = 0.562 = 0.25W$$

(a) $W = \dfrac{0.562}{0.25} = 2.248$ inches.

(b) The area of the conductor $= 846^2 = 715,716$ CM.

(c) The current-carrying capacity of the conductor $= 0.562 \times 1,000 = 562$ amps.

2-12 A certain 115-volt, 100-horsepower DC motor has an efficiency of 90% and requires a starting current of 150% times the full-load current. Determine (a) the size of fuses; (b) the copper requirements of the switch. Assume

OHM'S LAW AND OTHER ELECTRICAL FORMULAS
that one square inch of copper carries 1,000 amperes.
Motor current

$$I_M = \frac{hp \times 746}{E \times \text{efficiency}} = \frac{100 \times 746}{115 \times 0.9} = 721 \text{ amps}$$

(a) The amperage of the fuses is, therefore, 721 × 1.5 = 1,081.5, or 1,200.0 amperes, which is the rating of the closest manufactured fuse.

(b) Since the motor required a current of 721 amperes, the copper of each switch blade must be 721/1,000, or 0.721 square inch. Therefore, if ⅜-inch bus copper is used, its width must be:

$$W = \frac{0.721}{0.375} = 1.92 \text{ inches}$$

The mil-foot is a unit circular conductor that is one foot in length and one mil in diameter. The resistance of such a unit of copper has been found experimentally to be 10.37 ohms at 20°C.; this is normally thought of as 10.4 ohms (see Fig. 6).

A mil-foot of copper at 20°C. offers 10.4 ohms resistance; at 30°C., it is 11.2 ohms; at 40°C., it is 11.6 ohms; at 50°C., it is 11.8 ohms; at 60°C., it is 12.3 ohms; and at 70°C., it is 12.7 ohms. Thus, in voltage-drop calculations, 12 is generally used as the constant K, unless otherwise specified, to allow for higher temperatures and to afford some factor of safety.

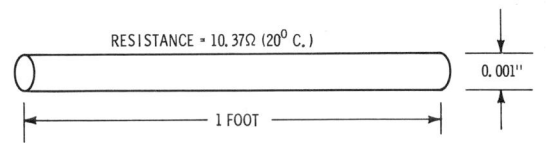

Fig. 6. Dimensions and resistance of one circular mil-foot of copper.

The resistance of a wire is directly proportional to its length and inversely proportional to its cross-sectional area. Therefore, if the resistance given in ohms of a mil-foot of wire is multiplied by the total length in feet (remember that there are practically always two wires, so if the distance is given in feet, multiply it by two to get the total resistance of both wires; this is a common error when working examinations) and divided

Ohm's Law and Other Electrical Formulas

by its cross-sectional area in circular mils, the result will be the total resistance of the wire in ohms. This is expressed as

$$R = \frac{K \times L \times 2}{A}$$

where,
  R is the resistance in ohms,
  K is the constant (12) for copper,
  L is the length in feet one way,
  A is the area in circular mils.

K for copper was given; so it will be well to give K for commercial aluminum:

20°C − 68°F − K = 17.39   50°C − 122°F − K = 19.73
30°C − 86°F − K = 18.73   60°C − 140°F − K = 20.56
40°C − 104°F − K = 19.40   70°C − 158°F − K = 21.23

K = 12 was used for copper, as it in most instances gives a safety factor in figuring voltage drop. Thus with a correction factor of 1.672 for aluminum, as opposed to copper, it would be well to use a *"K"* factor of 20 for aluminum.

Should you have other conditions, *"K"* factors have been given for both copper and aluminum at various degrees Celsius. To change Celsius to Fahrenheit, use the following formula:

Degrees C × 9/5 + 32 = Degrees F

Thus:

30°C × 9/5 + 32 =
270/5 + 32 = 54 + 32 = 86°F

2-13 What is the resistance of a 500-foot line of No. 4 copper wire?

From Table 1, No. 4 wire has a cross-sectional area of 41,740 CM. Therefore

$$R = \frac{12 \times 500 \times 2}{41,740} = \frac{12,000}{41,740} = 0.29 \text{ ohm}$$

Table 1 gives a value of 0.2480 ohm for 1,000 feet of No. 4 wire at 20°C. The difference is that we used a K of 12 instead of 10.4; 20°C. is 68°F., and the 12 value is for slightly over 50°C., or 122°F.

Ohm's Law and Other Electrical Formulas

2-14 Suppose it is desired to have a copper wire of 0.5 ohm resistance whose total length is 2,000 feet, or 1,000 feet one way. What must be its cross-sectional area? What size wire is necessary?

$$A = K\frac{L}{R} = \frac{12 \times 2,000}{0.5} = 48,000 \text{ CM}$$

According to Table 1, No. 4 wire has a cross-sectional area of 41,740 CM, and No. 3 wire has an area of 52,620 CM, so we would use the larger size No. 3 wire. Also notice that the equation did not use the factor 2, because the 2,000-foot length was the total and not just the length of one wire.

2-15 If the resistance of a copper wire whose diameter is 1/8 inch is measured as 0.125 ohm, what is the length of the wire?

$$L = \frac{RA}{K} = \frac{0.125 \times 125^2}{12} = 163 \text{ ft.}$$

Here again this is the total length; one way would be 81.5 feet.

2-16 A copper line that is 5 miles in length has a diameter of 0.25 inch. Calculate: (a) the diameter of the wire in mils; (b) the area of the wire in circular mils; (c) the weight in pounds; (d) the resistance at 50°C.

(Assume that the weight in pounds per cubic inch is 0.321.)
(a) The diameter in mils = $1,000 \times 0.25 = 250$ mils.
(b) The area in CM = $250^2 = 62,500$ CM.
(c) Cross-sectional area = $0.7854 \times D^2 = 0.7854 \times 0.25^2$
= 0.0491 sq. in.
Length of wire = $5,280 \times 5 \times 2 \times 12 = 633,600$ inches
Weight of wire = $0.0491 \times 633,600 \times 0.321 = 9,986.23$
or about 10,000 lbs.

(d) $$R = \frac{K \times 2 \times (5 \times 5,280)}{62,500} = \frac{12 \times 2 \times 26,400}{62,500}$$
$$= 10.13 \text{ ohms}$$

65

OHM'S LAW AND OTHER ELECTRICAL FORMULAS

SKIN EFFECT

When alternating current flows through a conductor, an inductive effect occurs, which tends to force the current to the surface of the conductor. This produces a voltage loss and also affects the current-carrying capacity of the conductor. For open wires or wires in nonmetallic-sheathed cable, this effect is neglected until the No. 0 wire size is reached. In metallic-sheathed cables and metallic raceways, the skin effect is neglected until size No. 2 is reached. At these points, there are multiplying factors for conversion from DC resistance to AC resistance (Table 2). Note that there is a different factor for aluminum than for copper cables.

2-17 **The DC resistance of a length of 250,000-CM copper cable in conduit (rigid metal) was found to be 0.05 ohm. What would its AC resistance be?**

From Table 2, the multiplying factor is found to be 1.06; therefore

$$R_{AC} = 0.05 \text{ ohm} \times 1.06 = 0.053 \text{ ohm}$$

VOLTAGE DROP CALCULATIONS

We have discussed the methods for finding the resistance of wire. By the use of Ohm's law, we can now find the voltage drop for circuit loads. Use the form

$$E = I \times R$$

Under Section 215-2(c) of the NEC, we find the prescribed maximum allowable percent of voltage drop permitted for feeders and for feeders and branch circuits. In Section 210-19(a) of the NEC, the maximum allowable percent of voltage drop for branch circuits is given.

2-18 **What is the percentage of allowable voltage drop for feeders that are used for power and heating loads?**
Maximum of 3%.

2-19 **What is the percentage of allowable voltage drop for feeders that are used for lighting loads?**
Maximum of 1% to 3%.

Ohm's Law and Other Electrical Formulas
Table 2. Multiplying Factors for Converting DC Resistance to 60-Cycle AC Resistance

Size	For Nonmetallic-Sheathed Cables in Air or Nonmetallic Conduit (Copper)	For Nonmetallic-Sheathed Cables in Air or Nonmetallic Conduit (Aluminum)	For Metallic-Sheathed Cables or All Cables in Metallic Raceways (Copper)	For Metallic-Sheathed Cables or All Cables in Metallic Raceways (Aluminum)
Up to 3 AWG	1.	1.	1.	1.
2	1.	1.	1.01	1.00
1	1.	1.	1.01	1.00
0	1.001	1.000	1.02	1.00
00	1.001	1.001	1.03	1.00
000	1.002	1.001	1.04	1.01
0000	1.004	1.002	1.05	1.01
250 MCM	1.005	1.002	1.06	1.02
300 MCM	1.006	1.003	1.07	1.02
350 MCM	1.009	1.004	1.08	1.03
400 MCM	1.011	1.005	1.10	1.04
500 MCM	1.018	1.007	1.13	1.06
600 MCM	1.025	1.010	1.16	1.08
700 MCM	1.034	1.013	1.19	1.11
750 MCM	1.039	1.015	1.21	1.12
800 MCM	1.044	1.017	1.22	1.14
1000 MCM	1.067	1.026	1.30	1.19
1250 MCM	1.102	1.040	1.41	1.27
1500 MCM	1.142	1.058	1.53	1.36
1750 MCM	1.185	1.079	1.67	1.46
2000 MCM	1.233	1.100	1.82	1.56

2-20 What is the percentage of allowable voltage drop for combined lighting, heating, and power loads?

Maximum of 1% to 3% for feeders and 5% for feeders and branch circuits. The following is the formula used for voltage-drop calculations. Since

$$R = \frac{K \times L \times 2}{A}$$

then

$$E_d = I \times R = \frac{K \times 2L \times I}{A}$$

OHM'S LAW AND OTHER ELECTRICAL FORMULAS

where,
  E_d is the voltage drop of the circuit,
  $2L$ is the total length of the wire,
  K is a constant (12),
  I is the current, in amperes, of the circuit,
  A is the area, in circular mils, of the wire in the circuit.

By transposing the formula above, we can determine the circular mil area of a wire for a specified voltage drop.

$$A = \frac{K \times 2L \times I}{E_d}$$

2-21 A certain motor pulls 22 amps at 230 volts, and the feeder circuit is 150 feet in length. If No. 10 copper wire is desired, what would be the voltage drop? Would No. 10 wire be permissible to use?

$$E_d = \frac{12 \times 2 \times 150 \times 22}{10,380}$$

$$E_d = \frac{79,200}{10,380} = 7.63 \ volts$$

However, 230 × 0.03 = 6.90 volts, which is the voltage drop permissible under section 215-3 of the NEC, so No. 10 wire would not be large enough.

2-22 In question 2-21, a voltage drop of 7.63 volts was calculated; however, the maximum permissible voltage drop is 6.9 volts. What size wire would have to be used?

According to Table 1, No. 10 wire has an area 10,380 CM, and No. 9 wire has an area of 13,090 CM. Therefore, No. 9 would be the proper size, except that you cannot purchase No. 9 wire; you will have to use No. 8 wire, which has an area of 16,510 CM.

2-23 If No. 8 copper wire were used in question 2-22, what would the voltage drop be? What percentage would this drop be?

$$E_d = \frac{12 \times 2 \times 150 \times 22}{16,509} = \frac{79,200}{16,509} = 4.80 \ volts$$

$$\frac{4.79}{230} \times 100 = 2.09 \ \%$$

OHM'S LAW AND OTHER ELECTRICAL FORMULAS

$$A = \frac{K \times 2L \times I}{E_d}$$

$$A = \frac{12 \times 2 \times 150 \times 22}{6.9}$$

$$A = \frac{79{,}200}{6.9} = 11{,}478.3 \text{ CM}$$

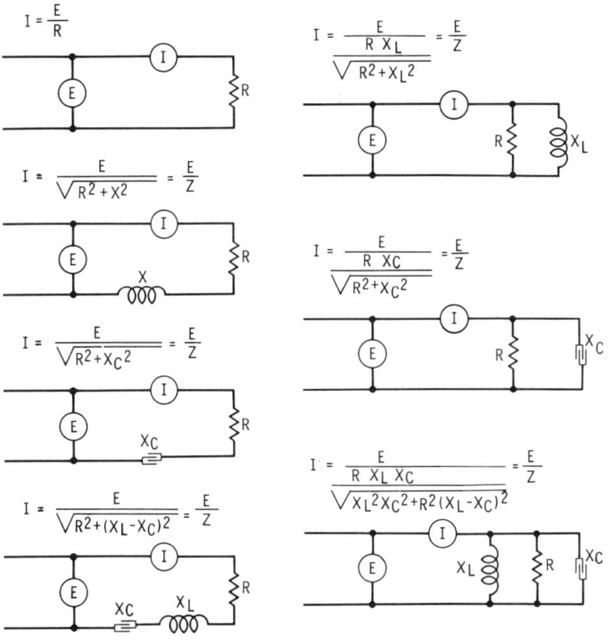

$I = \frac{E}{R}$

$I = \frac{E}{\sqrt{R^2 + X^2}} = \frac{E}{Z}$

$I = \frac{E}{\sqrt{R^2 + X_C^2}} = \frac{E}{Z}$

$I = \frac{E}{\sqrt{R^2 + (X_L - X_C)^2}} = \frac{E}{Z}$

$I = \frac{E}{\frac{R \; X_L}{\sqrt{R^2 + X_L^2}}} = \frac{E}{Z}$

$I = \frac{E}{\frac{R \; X_C}{\sqrt{R^2 + X_C^2}}} = \frac{E}{Z}$

$I = \frac{E}{\frac{R \; X_L \; X_C}{\sqrt{X_L^2 X_C^2 + R^2 (X_L - X_C)^2}}} = \frac{E}{Z}$

Fig. 7. Fundamental forms of AC circuits, with the method of determining current when voltage and impedance are known.

FORMULAS FOR DETERMINING ALTERNATING CURRENT IN ALTERNATING-CURRENT CIRCUITS

In the foregoing formulas of Fig. 7:

R is the resistance in ohms

X_L is the inductive reactance in ohms $= 2\pi f L$

X_C is the capacitive reactance in ohms $= \dfrac{1}{2\pi f C}$

f is the frequency

Ohm's Law and Other Electrical Formulas

L is the inductance in henrys
C is the capacity in farads
Z is the impedance in ohms
I is the current in amperes
E is the pressure in volts

FORMULAS FOR COMBINING RESISTANCE AND REACTANCE

In the following formulas of Fig. 8:
R is the resistance in ohms
X_L is the inductive reactance in ohms $= 2\pi f L$
X_C is the capacitive reactance in ohms $= \dfrac{1}{2\pi f C}$
f is the frequency
L is the inductance in henrys
C is the capacity in farads
Z is the impedance in ohms
I is the current in amperes
E is the pressure in volts

2-24 A coil has a resistance of 10 ohms and an inductance of 0.1 henry. If the frequency of the source is 60 cycles, what is the voltage necessary to cause a current of 2 amperes to flow through the coil?

With reference to the previously derived formula,

$$E_R = IR = 2 \times 10 = 20 \text{ volts}$$
$$X_L = 2\pi f L = 2\pi \times 60 \times 0.1 = 37.7 \text{ ohms}$$
$$E_L = IX_L = 2 \times 37.7 = 75.4 \text{ volts}$$

The applied voltage must therefore be

$$E = \sqrt{E_R^2 + E_L^2} = \sqrt{20^2 + 75.4^2} = 78 \text{ volts}$$

2-25. A coil with a negligible resistance requires 3 amperes when it is connected to a 180-volt, 60-cycle supply. What is the inductance of the coil?

$$X_L = \frac{E}{I} = \frac{180}{3} = 60 \text{ ohms}$$

OHM'S LAW AND OTHER ELECTRICAL FORMULAS

and
$$X_L = 2\pi f L$$
Therefore
$$L = \frac{60}{2\pi \times 60} = 0.159 \text{ henry}$$

2-26 An alternating current of 15 amperes with a frequency of 60 cycles is supplied to a circuit containing a resistance of 5 ohms and an inductance of 15 millihenrys. What is the applied voltage?

$$X_L = 2\pi f L = 2\pi \times 60 \times 0.015 = 5.65 \text{ ohms}$$
$$Z = \sqrt{R^2 + X_L^2} = \sqrt{5^2 + 5.65^2} = 7.54 \text{ ohms}$$
$$E = IZ = 15 \times 7.54 = 113.1 \text{ volts}$$

2-27 A coil contains 5 ohms resistance and 0.04 henry inductance. The voltage and frequency of the source are 100 volts at 60 cycles. Find (a) the impedance of the coil; (b) the current through the coil; (c) the voltage drop across the inductance; (d) the voltage drop across the resistance.

$$X_L = 2\pi f L = 2\pi \times 60 \times 0.04 = 15 \text{ ohms}$$

(a) $\quad Z = \sqrt{5^2 + 15^2} = \sqrt{250} = 15.8 \text{ ohms}$

(b) $\quad I = \dfrac{E}{Z} = \dfrac{100}{\sqrt{5^2 + 15^2}} = 6.3 \text{ amperes}$

(c) $\quad E_L = IX_L = 6.3 \times 15 = 94.5 \text{ volts}$

(d) $\quad E_R = IR = 6.3 \times 5 = 31.5 \text{ volts}$

2-28 An alternating-current circuit contains 10 ohms resistance in series with a capacitance of 40 microfarads. The voltage and frequency of the source are 120 volts at 60 cycles. Find (a) the current in the circuit; (b) the voltage drop across the resistance; (c) the voltage drop across the capacitance; (d) the power factor; (e) the power loss.

$$X_C = \frac{1}{2\pi f C} = \frac{1}{2\pi \times 60 \times 0.00004} = 66.67 \text{ ohms}$$
$$Z = \sqrt{10^2 + 66.7^2} = 67 \text{ ohms}$$

Ohm's Law and Other Electrical Formulas

(a) $I = \dfrac{E}{Z} = \dfrac{120}{67} = 1.8$ amperes

(b) $E_R = IR = 1.8 \times 10 = 18$ volts

(c) $E_C = IX_C = 1.8 \times 66.3 = 119.3$ volts

(d) $\cos \phi = \dfrac{R}{Z} = \dfrac{10}{67} = 0.15$ or 15%

(e) $P = E \times I \times \cos \phi = 120 \times 1.8 \times 0.15 = 32.4$ watts

2-29 A coil of 3 ohms resistance and 20 millihenrys inductance is connected in series with a capacitance of 400 microfarads. If the voltage and frequency are 120 volts at 60 cycles, find (a) the impedance of the circuit; (b) the current in the circuit; (c) the power loss; (d) the power factor.

$$X_L = 2\pi fL = 2\pi \times 60 \times 0.020 = 7.54 \text{ ohms}$$

$$X_C = \dfrac{1}{2\pi \times 60 \times 0.0004} = 6.67 \text{ ohms}$$

(a) $Z = \sqrt{R^2 + (X_L - X_C)^2} = \sqrt{3^2 + (7.54 - 6.67)^2}$
$= 3.14$ ohms

(b) $I = \dfrac{E}{Z} = \dfrac{120}{3.14} = 38.2$ amperes

(c) $P = I^2 R = 38.2^2 \times 3 = 4378$ watts

(d) $\cos \phi = \dfrac{R}{Z} = \dfrac{3}{3.14} = 0.955$ or 96% (approximately)

As a check for the power loss, in part (c), use the information obtained in part (d), the power factor. Then

$$P = EI \cos \phi = 120 \times 38.2 \times 0.958 = 4378 \text{ watts}$$

2-30 A certain series circuit has a resistance of 10 ohms, a capacitance of 0.0003 farad, and an inductance of 0.03

Ohm's Law and Other Electrical Formulas

henry. If a 60-cycle, 230-volt emf is applied to this circuit, find (a) the current through the circuit; (b) the power factor; (c) the power consumption.

$$X_L = 2\pi fL = 2\pi \times 60 \times 0.03 = 11.30 \text{ ohms}$$

$$X_C = \frac{1}{2\pi fC} = \frac{1}{2\pi \times 60 \times 0.0003} = 8.85 \text{ ohms}$$

$$Z = \sqrt{R^2 + (X_L - X_C)^2} = \sqrt{10^2 + (11.30 - 8.85)^2}$$
$$= 10.2 \text{ ohms}$$

(a) $$I = \frac{E}{Z} = \frac{230}{10.2} = 22.5 \text{ amperes}$$

(b) $$\cos \phi = \frac{10}{10.2} = 0.98 \text{ or } 98\%$$

(c) $$P = I^2R = 22.5^2 \times 10 = 5.06 \text{ kw}$$

2-31 The circuit in Fig. 9 contains a resistance of 30 ohms and a capacitance of 125 microfarads. If an alternating current of 8 amperes at a frequency of 60 cycles is flowing in the circuit, find (a) the voltage drop across the resistance; (b) the voltage drop across the capacitance; (c) the voltage applied across the circuit.

$$X_C = \frac{1}{2\pi fC} = \frac{1}{2\pi \times 60 \times 0.000125} = 21.2 \text{ ohms}$$
$$Z = \sqrt{R^2 + X_C^2} = \sqrt{30^2 + 21.2^2} = 36.7 \text{ ohms}$$

(a) $E_R = IR = 8 \times 30 = 240 \text{ volts}$
(b) $E_C = IX_C = 8 \times 21.2 = 170 \text{ volts}$
(c) $E = IZ = 8 \times 36.7 = 294 \text{ volts}$

2-32 A resistance of 15 ohms is connected in series with a capacitance of 50 microfarads. If the voltage of the source is 120 volts at 60 cycles, find (a) the amount of current in the circuit; (b) the voltage drop across the resistance; (c) the voltage drop across the capacitance;

Ohm's Law and Other Electrical Formulas

(d) the angular difference between the current and the applied voltage; (e) the power loss in the circuit.

$$X_C = \frac{1}{2\pi fC} = \frac{1}{2\pi \times 60 \times 0.00005} = 53.1 \text{ ohms}$$

$$Z = \sqrt{R^2 + X_C^2} = \sqrt{15^2 + 53.1^2} = 55.2 \text{ ohms}$$

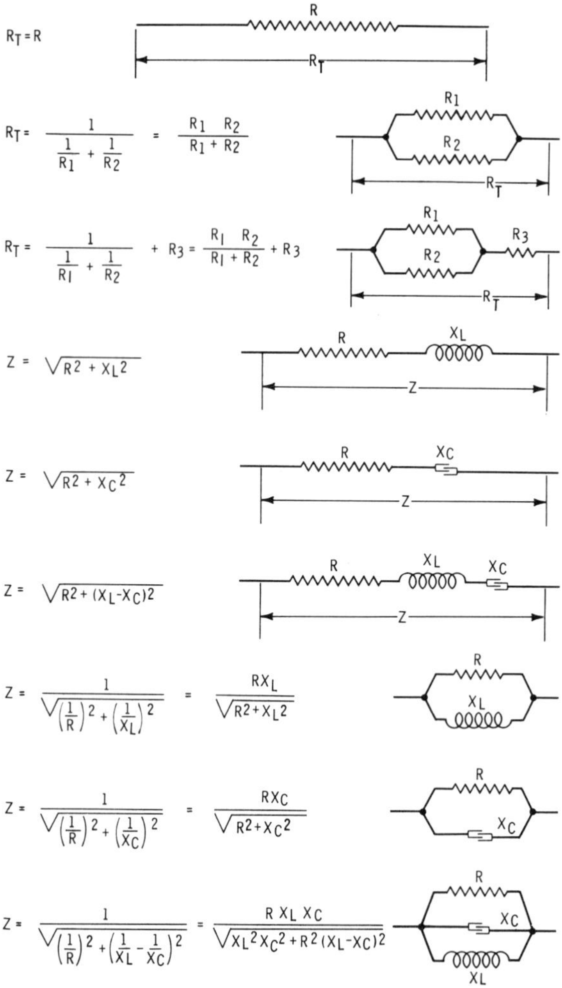

$R_T = R$

$$R_T = \frac{1}{\frac{1}{R_1} + \frac{1}{R_2}} = \frac{R_1 R_2}{R_1 + R_2}$$

$$R_T = \frac{1}{\frac{1}{R_1} + \frac{1}{R_2}} + R_3 = \frac{R_1 R_2}{R_1 + R_2} + R_3$$

$$Z = \sqrt{R^2 + X_L^2}$$

$$Z = \sqrt{R^2 + X_C^2}$$

$$Z = \sqrt{R^2 + (X_L - X_C)^2}$$

$$Z = \frac{1}{\sqrt{\left(\frac{1}{R}\right)^2 + \left(\frac{1}{X_L}\right)^2}} = \frac{RX_L}{\sqrt{R^2 + X_L^2}}$$

$$Z = \frac{1}{\sqrt{\left(\frac{1}{R}\right)^2 + \left(\frac{1}{X_C}\right)^2}} = \frac{RX_C}{\sqrt{R^2 + X_C^2}}$$

$$Z = \frac{1}{\sqrt{\left(\frac{1}{R}\right)^2 + \left(\frac{1}{X_L} - \frac{1}{X_C}\right)^2}} = \frac{R X_L X_C}{\sqrt{X_L^2 X_C^2 + R^2 (X_L - X_C)^2}}$$

Fig. 8. Methods of impedance determination in AC circuits when resistance, inductive reactance, and capacitive reactance, in ohms, are known.

OHM'S LAW AND OTHER ELECTRICAL FORMULAS

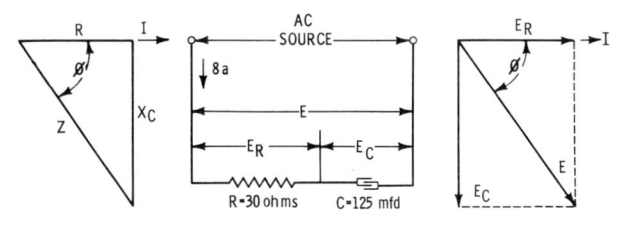

Fig. 9. Resistance and capacitance in series, with appropriate vector diagrams.

(a) $\quad I = \dfrac{E}{Z} = \dfrac{120}{55.2} = 2.17 \; amperes$

(b) $\quad E_R = IR = 2.17 \times 15 = 32.6 \; volts$

(c) $\quad E_C = IX_C = 2.17 \times 53.1 = 115.2 \; volts$

(d) $\quad \cos \phi = \dfrac{E_R}{E} = \dfrac{32.6}{120} = 0.272, \; and \; \phi = 74.2°$

(e) $\quad P = I^2R = 2.17^2 \times 15 = 71 \; watts$

2-33 A resistance of 20 ohms and a capacitance of 100 microfarads are connected in series across a 200-volt, 50-cycle AC supply; find (a) the current in the circuit; (b) the potential drop across the resistance; (c) the potential drop across the capacitance; (d) the phase difference between the current and the applied voltage; (e) the power consumed; (f) the power factor.

$$X_C = \dfrac{1}{2\pi fC} = \dfrac{1}{2\pi \times 50 \times 0.0001} = 31.8 \; ohms$$

$$Z = \sqrt{R^2 + X_C^2} = \sqrt{20^2 + 31.8^2} = 37.6 \; ohms$$

(a) $\quad I = \dfrac{E}{Z} = \dfrac{200}{37.6} = 5.32 \; amperes$

(b) $\quad E_R = IR = 5.32 \times 20 = 106.4 \; volts$

(c) $\quad E_C = IX_C = 5.32 \times 31.8 = 169.2 \; volts$

(d) $\quad \cos \phi = \dfrac{R}{Z} = \dfrac{20}{37.6} = 0.532, \; and \; \phi = 57.9°$

(e) $\quad P = I^2R = 5.32^2 \times 20 = 566 \; watts$

(f) $\quad \cos \phi = 0.532 \; or \; 53.2\%$

Ohm's Law and Other Electrical Formulas

2-34 What is the total capacitance of four parallel capacitors that are 10, 15, 25 and 30 microfarads, respectively?

$$C = C_1 + C_2 + C_3 + C_4$$
$$C = 10 + 15 + 25 + 30 = 80 \text{ microfarads}$$

2-35 A certain coil (shown in Fig. 10) with a resistance of 5 ohms and an inductance of 0.01 henry is connected in

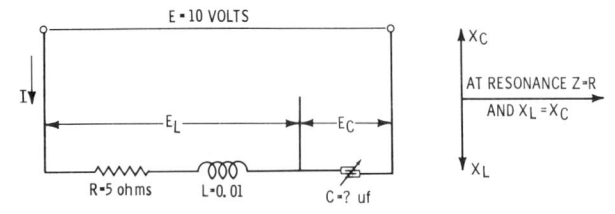

Fig. 10. Resistance, inductance, and capacitance in series, with a vector diagram illustrating their relationship to each other.

series with a capacitor across a 10-volt supply, which has a frequency of 800 cycles per second. Find (a) the capacitance that will produce resonance; (b) the corresponding value of the current; (c) the potential drop across the coil; (d) the potential drop across the capacitor; (e) the power factor of the circuit; (f) the power consumption.

The inductive reactance of the coil is

$$X_L = 2\pi \times 800 \times 0.01 = 50.24 \text{ ohms}$$

Since resonance occurs when $X_L = X_C$, X_C must also be equal to 50.24 ohms. Therefore,

(a) $$X_C = 50.24 = \frac{10^6}{2\pi \times 800 \times C}$$

$$C = \frac{10^4}{50.24^2}$$

$$C = 3.17 \text{ microfarads}$$

(b) At resonance, the current is

OHM'S LAW AND OTHER ELECTRICAL FORMULAS

$$I = \frac{E}{R} = \frac{10}{5} = 2 \text{ amperes}$$

(c) The potential drop across the coil is

$$E_L = IZ = 2\sqrt{5^2 + 50.24^2} = 2 \times 50.5 = 100.5 \text{ volts}$$

(d) The potential drop across the capacitor is

$$E_C = IX_C = 2 \times 50.24 = 100.5 \text{ volts}$$

(e) The power factor is

$$\cos \phi = \frac{R}{Z}$$

but since at resonance $Z = R$,

$$\cos \phi = \frac{R}{R} = 1, \text{ and } \phi = 0°$$

(f) The power consumed is

$$P = I^2R = 4 \times 5 = 20 \text{ watts}$$

2-36 The field winding of a shunt motor has a resistance of 110 ohms, and the emf applied to it is 220 volts. What is the amount of power expended in the field excitation?

The current through the field is

$$I_f = \frac{E_t}{R_f} = \frac{220}{110} = 2 \text{ amperes}$$

The power expended $= E_t I_f = 220 \times 2 = 440$ watts. The same results can also be obtained directly by using the equation

$$P_f = \frac{E_t^2}{R_f} = \frac{220^2}{110} = 440 \text{ watts}$$

2-37 A shunt motor whose armature resistance is 0.2 ohm and whose terminal voltage is 220 volts requires an armature current of 50 amperes and runs at 1,500 rpm when the field is fully excited. If the strength of the field is decreased and the amount of armature current is increased, both by 50%, at what speed will the motor run?

The expression for the counter emf of the motor is

77

OHM'S LAW AND OTHER ELECTRICAL FORMULAS

$$E_a = E_t - I_a R_a$$

and

$$E_{a1} = 220 - (50 \times 0.2) = 210 \text{ volts}$$

Similarly

$$E_{a2} = 220 - (75 \times 0.2) = 205 \text{ volts}$$

Also

$$E_a = N \phi K$$

and

$$\frac{E_{a1}}{E_{a2}} = \frac{N_1 \phi_1 K_1}{N_2 \phi_2 K_2}$$

Since the field is decreased by 50%, then

$$\phi_1 = 1.5 \phi_2, \text{ and } Z_1 = Z_2$$

It follows that

$$\frac{210}{205} = \frac{1,500 \times 1.5}{N_2}$$

$$N_2 = \frac{1,500 \times 205 \times 1.5}{210} = 2,196 \text{ rpm}$$

2-38 A 7.5 hp 220-volt interpole motor has armature and shunt-field resistances of 0.5 ohm and 200 ohms, respectively. The current input at 1,800 rpm under no-load conditions is 3.5 amperes. What are the current and electromagnetic torque for a speed of 1,700 rpm?

Under no-load conditions (at 1,800 rpm),

$$I_a = I_L - I_f = 3.5 - \left(\frac{220}{200}\right) = 2.4 \text{ amperes}$$

$$\phi K_{NL} = \frac{E_t - (I_a R_a)}{N} = \frac{220 - (2.4 \times 0.5)}{1,800} = 0.1216$$

$$\phi K_{NL} = \phi K_{FL}$$

At 1,700 rpm,

$$I_a = \frac{E_t - (N \phi K)}{R_a} = \frac{220 - (1,700 \times 0.1216)}{0.5} = 26.6 \text{ amperes}$$

OHM'S LAW AND OTHER ELECTRICAL FORMULAS

$I_L = I_a + I_f = 26.6 + 1.1 = 27.7$ amperes

$T_e = 7.05 \phi K I_a = 7.05 \times 0.1216 \times 26.6 = 22.8$ ft.-lb.

2-39 The mechanical efficiency of a shunt motor whose armature and field resistances are 0.055 and 32 ohms, respectively, is to be tested by means of a rope brake. When turning at 1,400 rpm, the longitudinal pull on the 6-inch diameter pulley is 57 lbs. Simultaneous readings on the line voltmeter and ammeter are 105 volts and 35 amperes, respectively. Calculate (a) the counter emf developed; (b) the copper losses; (c) the efficiency.

$$I_a = I_L - I_f = 35 - \left(\frac{105}{32}\right) = 31.7 \text{ amperes}$$

(a) $E_a = E_t - (I_a R_a) = 105 - (31.7 \times 0.055)$
               $= 103.26$ volts

(b) $P_C = I_f^2 R_f + I_a^2 R_a = (3.3^2 \times 32) + (31.7^2 \times 0.055)$
            $= 404$ watts

(c)   Output $= \dfrac{2\pi \times 1,400 \times 3/12 \times 57}{33,000} = 3.8$ hp

      Input $= \dfrac{105 \times 35}{746} = 4.93$ hp

      $\eta_m = \dfrac{3.8}{4.93} = 0.771$ or 77%

2-40 A copper transmission line that is 1.5 miles in length is used to transmit 10 kilowatts from a 600-volt generating station. The voltage drop in the line is not to exceed 10% of the generating-station voltage. Calculate (a) the line current; (b) the resistance of the line; (c) the cross-sectional area of the wire.

(a) $I_L = \dfrac{10,000}{600} = 16.67$ amperes

The permissible voltage drop $= 600 \times 0.1 = 60$ volts.

(b) $R = \dfrac{60}{16.67} = 3.6$ ohms

OHM'S LAW AND OTHER ELECTRICAL FORMULAS

(c) $\quad 3.6 = \dfrac{10.4 \times 3 \times 5{,}280}{A}$

$$A = \dfrac{10.4 \times 3 \times 5{,}280}{3.6} = 45{,}760 \ CM$$

2-41 A trolley system ten miles long is fed by two substations that generate 600 volts and 560 volts, respectively. The resistance of the trolley wire and rail return is 0.3 ohm per mile. If a car located four miles from the 600-volt substation draws 200 amperes from the line, what is the voltage between the trolley collector and track? How much current is supplied by each substation?

With reference to Fig. 11, the following equation can be written:

$$I_1 + I_2 = 200 \ amperes \qquad (1)$$

That is, the arithmetical sum of the current drain from each substation must equal the current drawn by the trolley car. In a similar manner, the equations for the voltage drop in each branch of the trolley wire are

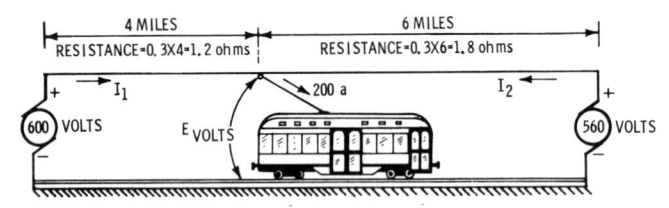

$$I_1(1.2) = 600 - E \qquad (2)$$
$$I_2(1.8) = 560 - E \qquad (3)$$

Fig. 11. Current and potential drop in a trolley-feeder system.

Subtracting equation (3) from equation (2),

$$40 = 1.2I_1 - 1.8I_2 \qquad (4)$$

According to equation (1),

$$I_1 = 200 - I_2$$

Therefore, equation (4) becomes

Ohm's Law and Other Electrical Formulas

$$40 = 1.2 (200 - I_2) - 1.8 I_2$$
$$I_2 = 66.67 \text{ amperes}$$
$$I_1 = 200 - 66.67 = 133.33 \text{ amperes}$$

By inserting the value of I_2 in equation (3), we obtain the voltage between the trolley collector and the track.

$$E = 560 - (1.8 \times 66.67) = 440 \text{ volts}$$

The same result can be obtained by inserting the value of I_1 in equation (2).

2-42 It is desired to supply power from a 220-volt source to points C and D in Fig. 12 by means of the feeder arrangement indicated. The motor at point C requires 120 amperes and the motor at point D requires 80 amperes. With the length of the wires as indicated and a maximum voltage drop of 10%, calculate (a) the cross-sectional area of feeder AB; (b) the cross-sectional area of feeder BC; (c) the cross-sectional area of feeder BD; (d) the power loss in each section.

The voltage drop across AC and AD is

$$E' = 220 \times 0.1 = 22 \text{ volts}$$

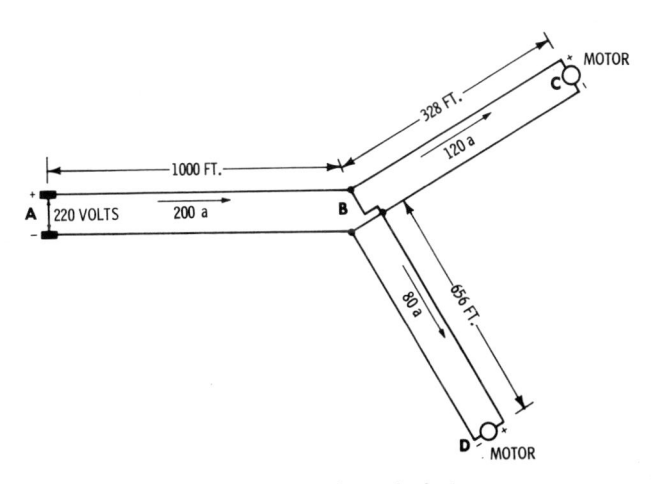

Fig. 12. Branch feeder calculations.

OHM'S LAW AND OTHER ELECTRICAL FORMULAS

To simplify our calculation, the voltage drop across BC and BD can be arbitrarily set at 10 volts. The voltage drop across AB is, therefore, $22 - 10$, or 12 volts.

(a) $\quad A = \left(\dfrac{10.4 \times 2}{12}\right) \times 200 \times 1{,}000 = 346{,}667 \text{ CM}$

(b) $\quad A = \left(\dfrac{10.4 \times 2}{10}\right) \times 120 \times 328 = 81{,}869 \text{ CM}$

(c) $\quad A = \left(\dfrac{10.4 \times 2}{10}\right) \times 80 \times 656 = 109{,}158 \text{ CM}$

(d) $\quad P_{AB} = 200 \times 12 = 2{,}400 \text{ watts}$
$\quad\quad P_{BC} = 120 \times 10 = 1{,}200 \text{ watts}$
$\quad\quad P_{BD} = 80 \times 10 = 800 \text{ watts}$

2-43 The motor illustrated in Fig. 13 is located at a distance of 500 feet from the generator and requires 40 amperes at 220 volts. Number 4 AWG wire is used; calculate (a) the voltage at the generator; (b) the voltage-drop percentage in the line; (c) the power loss in the line; (d) the power-loss percentage; (e) the cost of power losses per year. (Assume that the motor operates 8 hours per day, 300 days per year, at a cost of 3 cents per kilowatt-hour.)

With reference to Table 1, the cross-sectional area of No. 4 wire = 41,740 CM.

$$R = \dfrac{10.4 \times 1{,}000}{41{,}740} = 0.25 \text{ ohm}$$

(a) $\quad E_G = 220 + (40 \times 0.25) = 230 \text{ volts}$

(b) $\quad \dfrac{(230 - 220)\,100}{220} = 4.55\%$

(c) $\quad P_G - P_R = 40^2 \times 0.25 = 400 \text{ watts}$

(d) $\quad \dfrac{(P_G - P_R)\,100}{P_R} = \dfrac{400 \times 100}{40 \times 220} = 4.55\%$

(e) $\quad$ Yearly cost of power losses
$\quad\quad\quad = 0.4 \times 8 \times 300 \times 0.03 = \$28.80$

Ohm's Law and Other Electrical Formulas

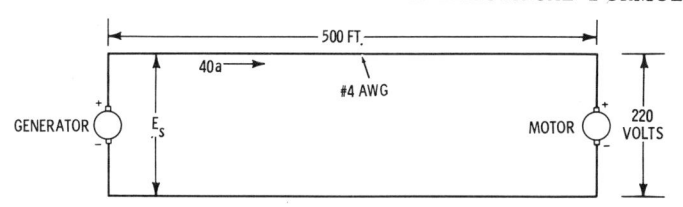

Fig. 13. Calculations for a 220-volt motor.

2-44 Energy is transmitted from a switchboard to the combined load shown in Fig. 14. The lamp group requires 20 amperes, and the motor requires 30 amperes from the line. Number 2 wire (resistance = 0.156 ohm per 1,000 ft.) is used throughout the circuit. Calculate (a) the power drawn by the lamps; (b) the power drawn by the motor; (c) the power loss in the line; (d) the total power supplied by the switchboard.

The resistance in line ABCD is

$$\frac{0.162 \times 200 \times 2}{1,000} = 0.065 \ ohm$$

Voltage across the lamps = 125 − (50 × 0.065) = 121.8 volts

(a) $\quad P = 20 \times 121.8 = 2.436 \ kw$

The resistance in line BEFC is

$$\frac{0.162 \times 100 \times 2}{1,000} = 0.0324 \ ohm$$

Voltage at the motor = 121.8 − (30 × 0.0324)
$\qquad\qquad\qquad\qquad\qquad\quad = 120.83 \ volts$

(b) $\quad P_M = 30 \times 120.83 = 3.63 \ kw$

Fig. 14. Voltage-drop calculations in a feeder circuit which is supplying a motor and lamp load.

OHM'S LAW AND OTHER ELECTRICAL FORMULAS

(c) $P_L = (50 \times 125) - (2{,}436 + 3{,}630) = 184$ watts
(d) $P_T = 125 \times 50 = 6.25$ kw

RELATIVE CONDUCTIVITY

2-45   What is the comparison of electrical conductivity between silver, copper, and aluminum?

Silver is the best conductor of electricity of the three and is considered as possessing 100% conductivity; copper is next and is considered as approximately 94% of the conductivity of silver; aluminum is considered last and has approximately 61% of the conductivity of silver.

2-46   What are the advantages and disadvantages of silver as a conductor?

The price of silver makes it generally prohibitive to use for a conductor; it is only used in special cases for its high conductivity where the price is of no consequence.

2-47   What are the advantages of copper as a conductor? What are its disadvantages?

Copper is plentiful, relatively cheap, a good conductor, and it has a high tensile strength. It is corrosive to some extent.

2-48   What are the advantages and disadvantages of aluminum as a conductor?

Aluminum is inexpensive, light weight, and readily available. It is more corrosive than copper, so its use is limited to some degree; it also has less tensile strength than copper.

CHAPTER 3

Power and Power Factor

The unit of *work* is the foot-pound. This is the amount of work done when a force of one pound acts through a distance of one foot. The amount of work done is equal to the force in pounds times the distance in feet, or

$$W \text{ (work)} = Force \times Distance$$

Thus, if an object weighing 10 pounds is lifted a distance of 4 feet, the work done is equal to 40 ft.-lb.

Time becomes involved when performing work, so we use the quantity known as *power*, which is the rate of doing work. Power is directly proportional to the amount of work done and is inversely proportional to the time in which the work is done. For example, more power is required if an object weighing 10 pounds is lifted through a distance of 4 feet in 1 minute than if the same 10-pound object is lifted a distance of 4 feet in 5 minutes. From this statement, we may arrive at a formula for power.

$$Power \text{ (foot-pounds per minute)} = \frac{Work\ done\ (foot\text{-}pounds)}{Time\ (minutes)}$$

The more common unit of power is the horsepower, which is equivalent to 33,000 ft.-lb. of work per minute. Remember this figure and its relationship to time and horsepower; it will be used quite often when working problems that deal with power.

In electricity, the unit of power is the watt; however, since the watt is a relatively small unit, the kilowatt is more commonly used for the unit of power. One kilowatt is equivalent to 1,000 watts.

A good working knowledge of the electrical formulas that are used to determine power is a *must* for the electrician. When using

Power and Power Factor

electrical formulas to determine power, it is a universal practice to use the following notations:

P is the power in watts,
I is the current in amperes,
R is the resistance in ohms,
E is the potential difference in volts.

Thus, the power P expended in a load resistance R when a current I flows due to a voltage pressure E can be found by the following relationships:

$$P = IE$$
$$P = \frac{E^2}{R}$$
$$P = I^2R$$

Remember, IR equals a potential, in volts, and I^2R equals power, in watts.

When dealing with large amounts of electrical power, it may be required that you be able to determine the cost of the power consumed. You will be dealing with units of kilowatts and kilowatt-hours (kwh), which means the number of kilowatts used per hour. Thus, 25 kwh is equivalent to 25 kilowatts used for 1 hour. To find the cost of an electrical-usage bill, the following formula is used

$$Cost = \frac{watts \times hours\ used \times rate\ per\ kwh}{1,000}$$

For example, an electric heater that draws 1,350 watts is used for 4 hours, and the cost of electricity for that particular location is 3 cents per kilowatt-hour. What is the cost of using the heater?

$$Cost = \frac{1,350 \times 4 \times 0.03}{1,000} = 16.2\ cents$$

Energy can be changed from one form to another but can never be destroyed. Therefore, we may readily change electrical power into mechanical power; the converse is also true. The usual method of referring to mechanical power is in terms of units of horsepower; one horsepower is equal to 746 watts. This equality is valid when considering that the equipment used to produce one horsepower operates at 100% efficiency, which, of course, is not possible, since

POWER AND POWER FACTOR

there is always some power lost in the form of friction or other losses, which will be covered later in this text.

3-1 **A motor draws 50 amperes and is fed by a line that is 125 feet long; the line consists of No. 6 copper wire. What is the I^2R loss of the line?**

From Table 1 of Chapter 2, it is found that No. 6 copper has a resistance of 0.410 ohm per 1,000 feet at 25° C. The line is 125 feet long; therefore, the amount of wire used will be 2 × 125, or 250 feet. This is 25% of 1,000 feet, so the resistance of the wire will be 0.410 × 0.25, or 0.1025 ohm. Therefore,

$$P = I^2R = (50)^2 \times 0.1025 = 2,500 \times 0.1025 = 256.25 \ watts$$

3-2 **The line loss of question 3-1 is 246.5 watts. If the motor is operated for 100 hours, with the rate of electricity being 3 cents per kwh, what would be the cost of the I^2R loss of the line?**

$$Cost = \frac{256.25 \times 100 \times 0.03}{1,000} = 76.875 \ cents$$

3-3 **A 1-hp motor draws 1,000 watts. What is its efficiency?**

$$Efficiency = \frac{Output}{Input}$$

Therefore,

$$Efficiency = \frac{746}{1,000} = 0.746 \ or \ 74.6\%$$

3-4 **An electric iron draws 11 amperes at 120 volts. How much power is used by the iron?**

$$P = 11 \times 120 = 1,320 \ watts = 1.32 \ kw$$

3-5 **A motor must lift an elevator car weighing 2,000 pounds to a height of 1,000 feet in 4 minutes. (a) What is the theoretical size, in horsepower, of the motor required? (b) At 50% efficiency, what is the size, in horsepower, of the motor required?**

POWER AND POWER FACTOR

(a) $\quad W = 2{,}000 \times 1{,}000 = 2{,}000{,}000$ ft.-lb.

$$\frac{2{,}000{,}000}{4} = 500{,}000 \text{ ft.-lb. per minute}$$

$$\frac{500{,}000}{33{,}000} = 15.15 \text{ hp}$$

(b) $\quad Input = \dfrac{Output}{Efficiency} = \dfrac{15.15}{0.50} = 30.3$ hp

A 30-hp motor will carry this load nicely.

3-6 A lamp operating at 120 volts has a resistance of 240 ohms. What is the wattage of the lamp?

$$P = \frac{E^2}{R} = \frac{120^2}{240} = \frac{14{,}400}{240} = 60 \text{ watts}$$

3-7 What is the overall efficiency of a 5-hp motor that draws 20 amperes at 240 volts?

$$Input = 240 \times 20 = 4{,}800 \text{ watts}$$
$$Output = 5 \times 746 = 3{,}730 \text{ watts}$$
$$Efficiency = \frac{3{,}730}{4{,}800} = 0.777 \text{ or } 77.7\%$$

3-8 What is the cost of operating a 2-watt electric clock for one year at 2 cents per kwh?

$$Cost = \frac{2 \text{ watts} \times 24 \text{ hours} \times 365 \text{ days} \times 0.02}{1{,}000} = 35.04 \text{ cents}$$

3-9 The primary of a transformer draws 4 amperes at 7,200 volts. A reading at the secondary shows 110 amperes at 240 volts. What is the efficiency of the transformer at this load?

$$Efficiency = \frac{Output}{Input} = \frac{110 \times 240}{4 \times 7{,}200} = \frac{26{,}400}{28{,}800} = 91.67\%$$

3-10 What instrument is used to measure voltage?
   A voltmeter.

3-11 How is a voltmeter connected in a circuit (explain and illustrate)?

POWER AND POWER FACTOR

Fig. 1. Voltmeter connections in a circuit.

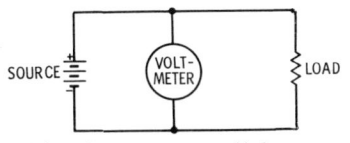

A voltmeter is always connected in shunt, or parallel, across the load or the source that you are measuring the voltage of. This is illustrated in Fig. 1.

3-12 With what instrument do you measure current?
An ammeter.

3-13 How is an ammeter connected in a circuit (explain and illustrate)?
An ammeter is connected in series with the circuit being tested. This is illustrated in Fig. 2.

Fig. 2. Ammeter connections in a circuit.

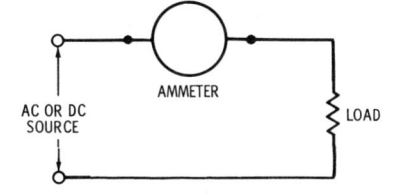

3-14 Give two methods of measuring large currents. Explain why they must be used.
When large currents are measured, it is impractical to have a meter constructed that is capable of carrying the large currents. Therefore, in AC or DC circuits, you use a shunt, which consists of a low resistance connected in series with the load and also connected in parallel with a high-resistance meter. The meter then receives only a small fraction of the current passing through the load. This method is illustrated in Fig. 3, shown below.

Fig. 3. One common method used to measure large current.

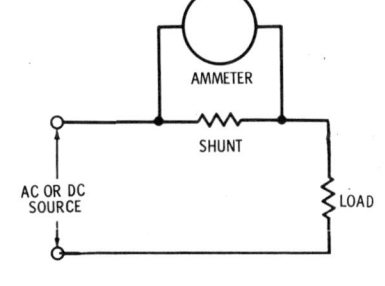

89

POWER AND POWER FACTOR

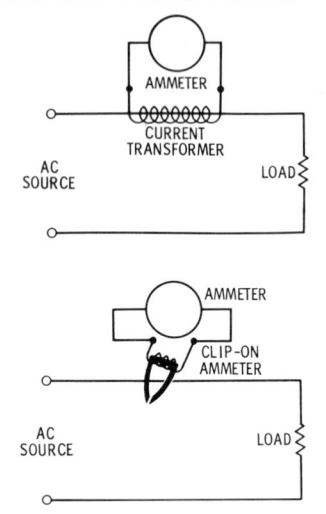

Fig. 4. Measurement of large currents in an AC circuit.

Another method for measuring large currents in an AC circuit makes use of a current transformer (Fig. 4A), which is described in the chapter on transformers. You may also use the wire that carries the load current as the primary of a current transformer in conjunction with a clip-on ammeter; the secondary of the transformer is incorporated in the meter, so the meter has no actual physical connection in the circuit (Fig. 4B).

3-15 When using an ammeter, what precautions must be taken?

Know whether the current is AC or DC, and use the appropriate ammeter. Be sure that the rating of the meter is large enough for the current being measured; this step will prevent the meter from being damaged.

3-16 What is a wattmeter?

A wattmeter is an instrument that is designed to indicate directly the active power in an electric circuit. It consists of a coil connected in series with the circuit, such as in an ammeter, and a coil connected in parallel with the circuit, such as in a voltmeter. Both coils actuate the same meter, thereby giving the measurement of both the current and the voltage affecting one meter, which may be calibrated in watts, kilowatts, or megawatts.

POWER AND POWER FACTOR

3-17 How is a wattmeter connected in a circuit? Illustrate (see Fig. 5).

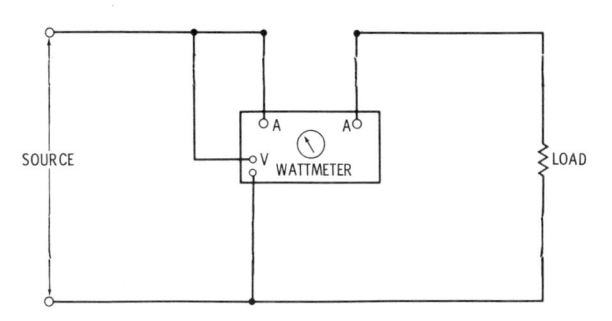

Fig. 5. Wattmeter connections in a circuit.

3-18 Can the principles used in DC circuits be applied to all AC circuits?

The fundamental principles of DC circuits may also apply to AC circuits that are strictly resistive in nature, such as incandescent lighting and heating loads.

3-19 What causes inductive reactance?

Inductive reactance is caused by the opposition to the flow of alternating current by the inductance of the circuit.

3-20 Give some examples of equipment that cause inductive reactance.

Motors, transformers, choke coils, relay coils, ballasts.

3-21 When only inductive reactance is present in an AC circuit, what happens to the current in relation to the voltage?

The current is said to *lag* behind the voltage.

3-22 What is the reason for current lagging the voltage in an inductive circuit?

In an AC circuit, the current is continually changing its direction of flow (60 times a second in a 60-hertz circuit). Any change of current value is opposed by the inductance within the circuit involved.

3-23 Draw sine waves of voltage and current in a circuit containing only inductive reactance when 60-hertz AC is applied to the circuit. (See Fig. 6.)

POWER AND POWER FACTOR

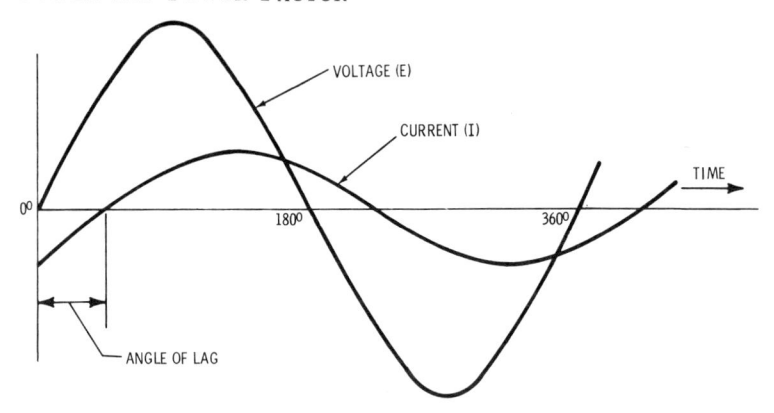

Fig. 6. Voltage and current sine waves in a circuit containing only inductive reactance, with a frequency of 60 hertz.

3-24 When only capacitive reactance is present in an AC circuit, what happens to the relationship that exists between the voltage and the current?

The current is said to *lead* the voltage.

3-25 Draw sine waves of current and voltage in a circuit containing only capacitive reactance when 60-hertz AC is applied to the circuit. (See Fig. 7.)

3-26 In a purely resistive circuit with an applied 60-hertz AC voltage, what is the relationship between the current and voltage?

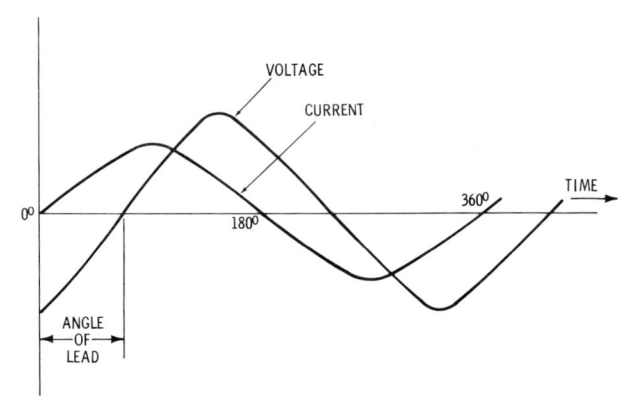

Fig. 7. Voltage and current sine waves in a circuit containing only capacitive reactance, with a frequency of 60 hertz.

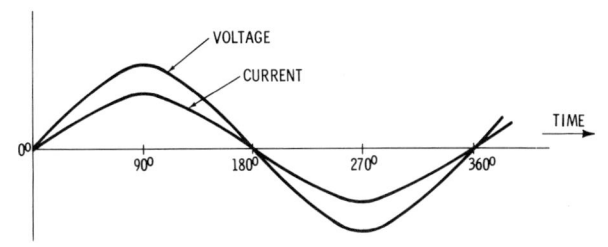

POWER AND POWER FACTOR

Fig. 8. Voltage-current relationship in a resistive circuit at a frequency of 60 hertz.

The current and voltage will be in phase, as illustrated in Fig. 8.

3-27 Is it possible to have a circuit with only inductive reactance?

This is an electrical impossibility because of the fact that metallic wire possesses resistance.

3-28 If it were possible to have only inductive reactance in an AC circuit, by what angle would the current lag the voltage?

The current would lag the voltage by 90°; that is, when the voltage was at its maximum value in one direction, the current would be zero and would be just getting ready to increase in the direction of maximum voltage.

3-29 If it were possible to have only capacitive reactance in an AC circuit, by what angle would the current lead the voltage?

The current would lead the voltage by 90°; that is, when the current was at its maximum value in one direction, the voltage would be zero and would be just getting ready to increase in the direction of maximum current.

3-30 Draw vectorially current and voltage 90° out of phase in an inductive circuit. (See Fig. 9.)

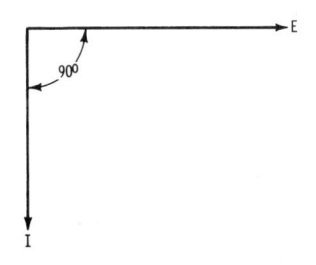

Fig. 9. Vector relationship between voltage and current in an inductive circuit.

POWER AND POWER FACTOR

3-31 Draw vectorially current and voltage 90° out of phase in a capacitive circuit. (See Fig. 10.)

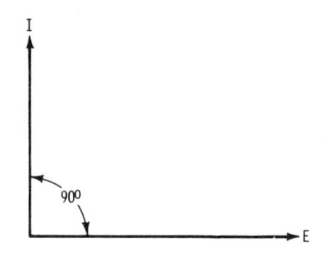

Fig. 10. Vector relationship between voltage and current in a capacitive circuit.

3-32 What is meant by power factor?

Power factor is the phase displacement of current and voltage in an AC circuit. The cosine of the phase angle of displacement is the power factor; the cosine is multiplied by 100 and is expressed as a percentage. The cosine of 90° is 0; therefore, the power factor is 0%. If the angle of displacement were 60°, the cosine of which is 0.500, the power factor would be 50%. This is true whether the current leads or lags the voltage.

3-33 How is power expressed in DC circuits and AC circuits that are purely resistive in nature?

In DC circuits and AC circuits that contain only resistance,

$$P\ (watts) = E \times I$$

$$P\ (kilowatts) = \frac{E \times I}{1,000}$$

3-34 How is power expressed in AC circuits that contain inductive and/or capacitive reactance?

$$VA\ (volt\text{-}amperes) = E \times I$$

$$KVA\ (kilovolt\text{-}amperes) = \frac{E \times I}{1,000}$$

$$P\ (watts) = E \times I \times Power\ Factor$$

3-35 In a 60-cycle AC circuit, if the voltage is 120 volts, the current is 12 amps, and the current lags the voltage by 60°, find (a) the power factor; (b) the power in volt-amperes (VA); (c) the power in watts.

POWER AND POWER FACTOR

(a) The cosine of 60° is 0.500; therefore, the power factor is 50%.
(b) $120 \times 12 = 1{,}440\ VA$, which is called the *apparent* power.
(c) $120 \times 12 \times 0.5 = 720\ watts$, which is called the *true* power.

3-36 Show vectorially a 12-amp line current lagging the voltage by 60°; indicate that the in-phase current is 50%, or 6 amps. (See Fig. 11.)

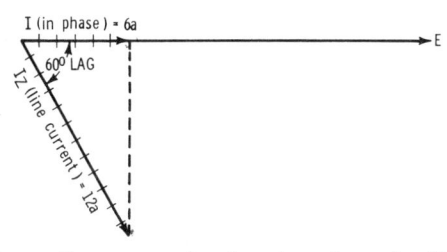

Fig. 11. A 12-amp line current lagging the voltage by 60°; the in-phase current is 50% of the line current, or 6 amps.

3-37 Why is a large power factor of great importance?

As can be seen in questions 3-35 and 3-36, there is an apparent power of 1,440 VA and a true power of 720 watts. There are also 12 amps of line current and 6 amps of in-phase, or effective current. This means that all equipment from the source of supply to the power-consumption device must be capable of handling a current of 12 amps, while actually the device is only utilizing a current of 6 amps. A 50% power factor was used intentionally to make the results more pronounced. The I^2R loss is based on the 12-amp current, while only 6 amps are really effective.

3-38 How can power factor be measured or determined?

There are two easy methods: (1) by the combined use of a voltmeter, ammeter and wattmeter (all AC instruments, of course), and (2) by the use of a power-factor meter.

3-39 How is the voltmeter-ammeter-wattmeter method used to determine the power factor?

The voltmeter, ammeter, and wattmeter are connected properly in the circuit. Then the readings of the three meters are simultaneously taken under the same load conditions. Finally, the following calculations are made.

Power and Power Factor

$$\text{Power Factor} = \frac{\text{True power (watts)}}{\text{Apparent power } (E \times I)}$$

$$\text{Power Factor} = \frac{w}{E \times I} = \frac{kw}{KVA}$$

3-40 What is a power-factor meter?

A power-factor meter is a wattmeter calibrated to read the power factor directly instead of in watts. It is connected in the circuit in the same manner as a wattmeter.

3-41 In a circuit that contains an inductance, such as a motor, will the current lead or lag the voltage? What steps can be taken to correct the power factor?

The current will lag the voltage. The power factor can be corrected by adding capacitance to the circuit or by using synchronous motors and overexciting their fields.

3-42 What is the ideal power-transmission condition, as far as the power factor is concerned?

Unity power factor, or a power factor of 1, which means that the current is in phase with the voltage.

3-43 Which is the best condition, a leading or lagging current?

They both have the same effect, so one is as good as the other. In an overall picture, there are usually more lagging currents than leading currents, since most of the loads used in the electrical field are of an inductive nature.

3-44 In an AC voltage or current, there are three values that are referred to. What are they?

The *maximum* value of current or voltage, the *effective* value of current or voltage, and the *average* value of current or voltage.

3-45 What is meant by the *maximum* AC voltage?

This is the maximum, or peak, voltage value of an AC sine wave.

3-46 What is meant by the *average* AC voltage?

This is the average of the voltages taken at all points on the sine wave.

3-47 What is meant by the *effective* AC voltage?

This is the value of the useful voltage that is indicated on a voltmeter; it is the voltage that is used in all normal calculations in electrical circuits.

POWER AND POWER FACTOR

3-48 What percentage of the peak voltage is the effective voltage?
70.7% of the peak value.

3-49 What percentage of the peak voltage is the average voltage?
63.7% of the peak voltage.

3-50 What is the effective voltage most commonly called?
The root-mean-square, or rms, voltage.

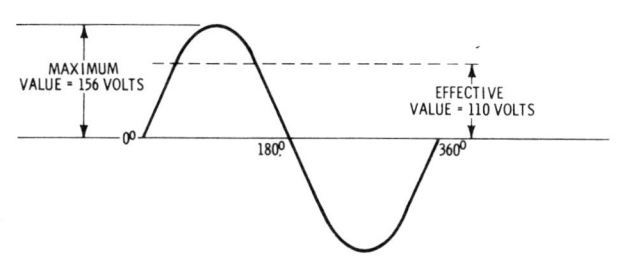

Fig. 12. The effective and peak values of a standard 110-volt line.

3-51 How could the rms voltage be arrived at vectorially?
An infinitesimal number of lines could be drawn from the base line to the one-half amplitude value of the sine wave. These voltage values would then be squared; the sum of the squares would be averaged (added together and divided by the number of squares), and the square root of this figure would be the rms value of the voltage in question.

3-52 Does the peak voltage have to be considered?
Yes. In design, the peak voltage must be considered, because this value is reached twice in every cycle.

3-53 Draw a sine wave showing the effective value and the peak value of a standard 110-volt AC line. (See Fig. 12.)

CHAPTER 4

Lighting

4-1 What is an identified conductor?
A white or natural gray conductor, which indicates that it is a grounded conductor.

4-2 May an identified conductor ever be used as a current-carrying conductor? Explain.
Yes, if the identified conductor is permanently rendered unidentified by painting or other effective means at each outlet where the conductors are visible and accessible (see NEC, Article 200-7, Exception 1). This is permissible where a two-wire cable having one black and one white wire is tapped from the outside wires of a three-wire circuit. It may also be used for a current-carrying conductor when used for single-pole, three- or four-way switch loops (see NEC, Article 200-7, Exception 2).

4-3 Draw a diagram of the proper connections for a two-wire cable, one white and one black, to supply a light from a single-pole switch. (See Fig. 1.)

4-4 When must the neutral, or identified, wire be provided with a switch?

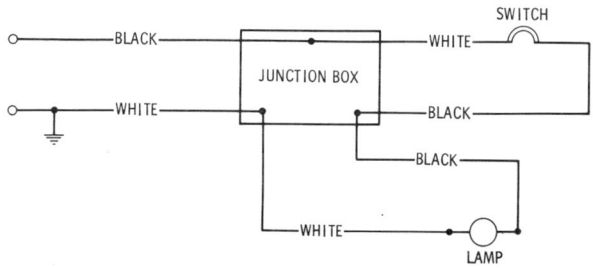

Fig. 1. A two-wire cable supplying a lamp from a single-pole switch.

LIGHTING

A lamp or pump circuit on a gasoline-dispensing island *must* provide both the neutral and the hot wires, which supply the light or the pump, with a switch (see NEC, Article 514-5).

4-5 Draw a three-way switch system and a lamp, using cable; show the colors and connections. (See Fig. 2.)

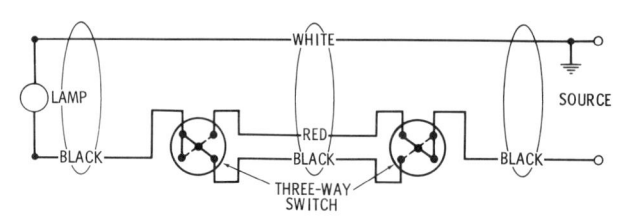

Fig. 2. A three-way switch system with a lamp.

4-6 Draw a circuit supplying a lamp that is controlled by two three-way switches and one four-way switch (use cable). (See Fig. 3.)
4-7 Draw a master-control lighting system. (See Fig. 4.)
4-8 Draw an electrolier switching circuit for controlling lights. (See Fig. 5.)
4-9 Draw an electrolier switching circuit to control three sets of lamps. (See Fig. 6.)

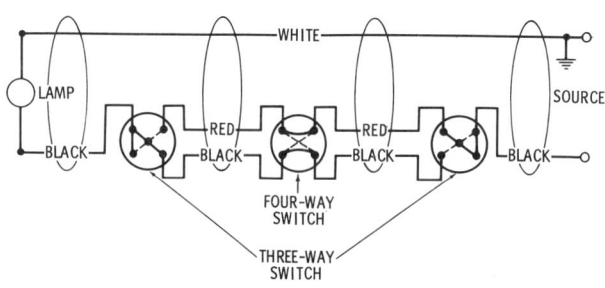

Fig. 3. A lamp circuit that is controlled by two three-way switches and one four-way switch.

4-10 What is a preheated-type fluorescent tube?

There are two external contacts at each end of a glass tube. Each set of contacts is connected to a specially treated tungsten filament. The inside of the tube is coated with a fluorescent powder. The type of powder used controls the color output of

LIGHTING

the tube. The tube is filled with inert gas, such as argon, and a small drop of mercury to facilitate starting.

4-11 What kind of light does the fluorescent bulb produce within the bulb itself?

Ultraviolet light.

A. One circuit with two groups of lamps.

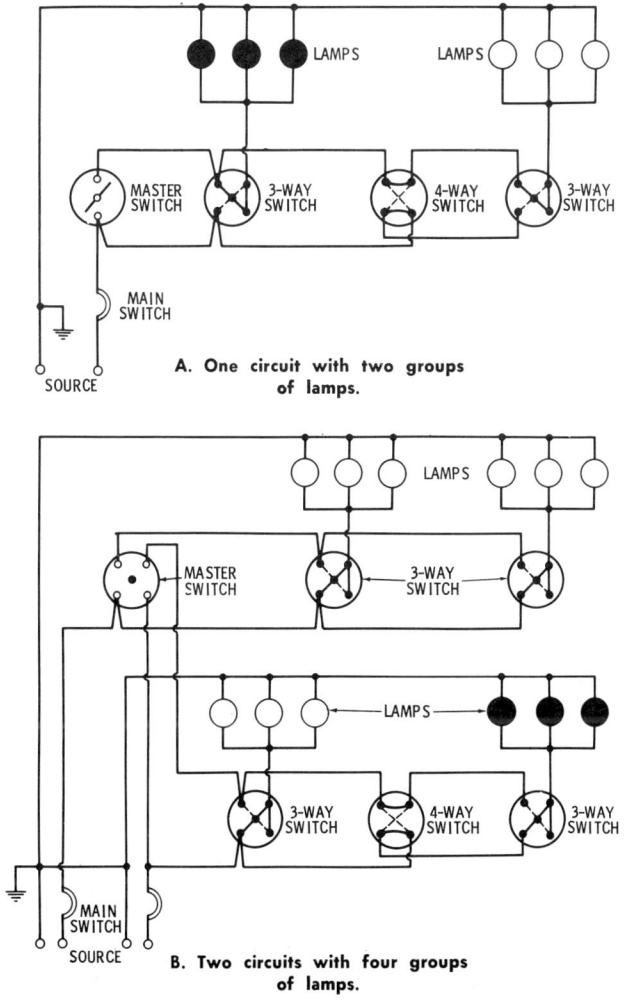

B. Two circuits with four groups of lamps.

Fig. 4. A master-control lighting system.

101

LIGHTING

4-12 Do the filaments stay lit during the operation of a fluorescent bulb? Explain.

No. They remain lit only at the start to vaporize the mercury; they are then shut off by the starter. Current is supplied to one contact on each end, thereby sustaining the mercury arc within the tube.

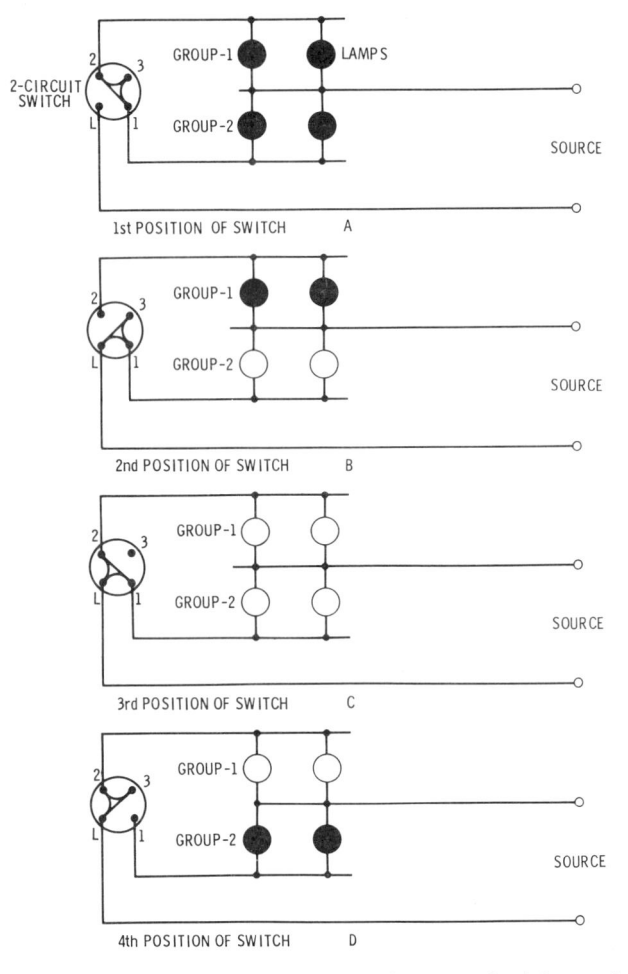

Fig. 5. An electrolier switch arrangement for control of lamp circuits.

102

LIGHTING

4-13 Draw a simple circuit of a fluorescent lamp and fixture. (See Fig. 7.)

4-14 Explain the action of the glow-type starter.

When the switch is closed, the high resistance in the glow bulb of the starter produces heat, which causes the bimetallic

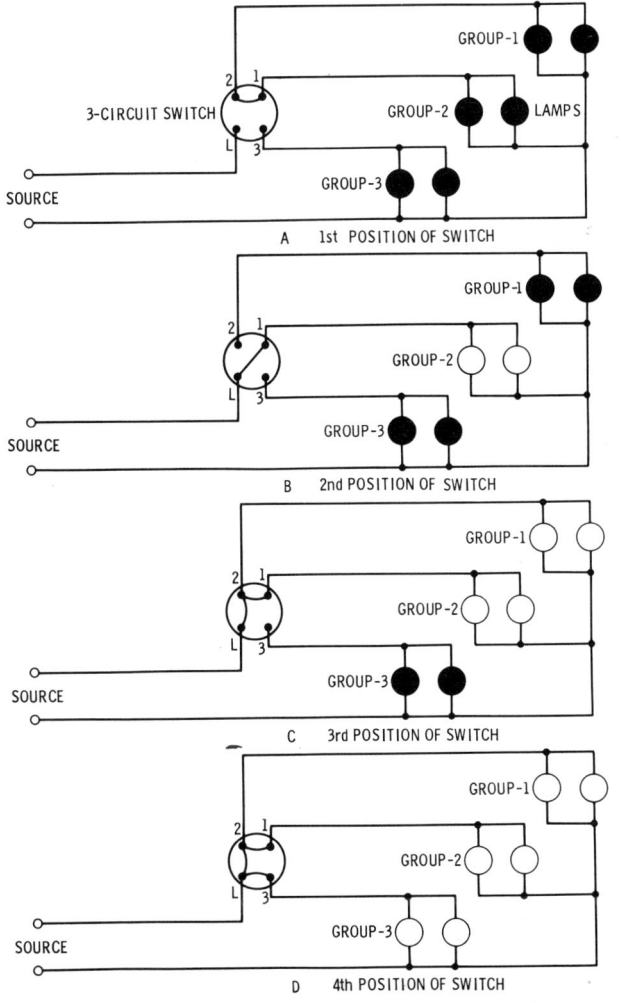

Fig. 6. An electrolier switch arrangement for control of three groups of lamps. The sequence of operation is depicted diagrammatically.

103

Lighting

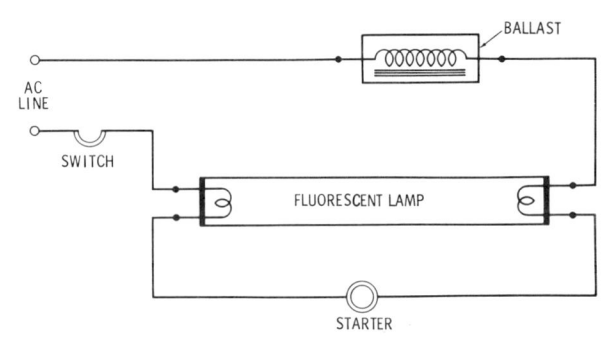

Fig. 7. A fluorescent lamp and fixture circuit.

U-shaped strip to close the contacts, thereby lighting the filaments in the fluorescent bulb. When the contacts close, the glow bulb cools and allows the contacts to open, thus disconnecting one side of the filaments. The arc within the fluorescent bulb is sustained without the filaments being heated, and the bulb lights.

4-15 Is the power factor of the circuit in Fig. 7 good or bad? Is the current leading or lagging the voltage?

The power factor is bad; the current is lagging the voltage because of the inductive reactance of the ballast.

4-16 How may the power factor of the circuit in Fig. 7 be corrected?

By the addition of capacitors in the ballast to counteract the lagging power factor.

4-17 What is the fixture in question 4-16 called?

A power-factor-corrected fixture.

4-18 What is a trigger-start fluorescent fixture?

This fixture uses a special trigger-start ballast that automatically preheats the filaments without the use of a starter.

4-19 Is a fluorescent fixture more efficient than an incandescent fixture?

Yes; however, no fixed efficiency can be quoted, since it varies with the size of the bulb, the ballast, etc. As a rule of thumb, a 40-w fluorescent bulb is generally considered to put out about the same amount of light as a 100-w incandescent bulb.

4-20 What is an instant-start Slimline fluorescent lamp?

This lamp has a single terminal at each end; the ballast is normally of the autotransformer type, which delivers a high

LIGHTING

voltage at the start and a normal voltage after the arc is established.

4-21 Does room temperature affect the operation of fluorescent lamps?

Yes, the normal fluorescent lamp is designed for operation at 50°F. or higher.

4-22 If the operating temperature is expected to be below 50°F., what measures should be taken as far as using fluorescent lamps?

Use special lamps and starters that are designed for lower temperatures.

4-23 Do fluorescent lamps produce a stroboscopic effect? Explain.

Yes; this effect is due to the fact that at 60 cycles, the current passes through zero 120 times a second.

4-24 How can the strobe effect be compensated for in a two-bulb system?

When the power factor is corrected, the capacitor is so connected that at the instant the current in one bulb is passing through zero, the current in the other bulb is not at zero, and the strobe effect goes unnoticed.

4-25 What is another way to minimize the strobe effect?

By using a higher frequency, such as 400 cycles.

4-26 Does a 400-cycle frequency have any other advantages?

Yes. Smaller and lighter ballasts can be used. This makes feasible the use of simple capacitance-type ballasts, which produce an overall gain in efficiency.

4-27 Does frequent starting and stopping of fluorescent lighting affect the bulb life?

The life of fluorescent bulbs is affected by starting and stopping. A bulb that is constantly left on will have a much longer life than one that is turned on and off often.

4-28 When lighting a football field with large incandescent bulbs, how can the light output of the bulbs be increased?

By using bulbs of a lower voltage rating than that of the source of supply. For example, by using a bulb that is rated at 105 volts on a 120-volt supply, the output can be increased roughly by 30%, but the life of the bulb will be cut by about 10%.

LIGHTING

4-29 How is the light output of lamps rated?
In lumens.

4-30 How is light measured?
In foot-candles.

4-31 What is a foot-candle?
The amount of direct light emitted by one international candle on a square foot of surface, every part of which is one foot away.

CHAPTER 5

Branch Circuits and Feeders

5-1 Is the NEC a law?
It is not a law, but it is adopted into laws that are established by governmental agencies; it may be adopted in its entirety, in part, or with amendments.

5-2 Who has the responsibility of Code interpretations?
The administrative authority that has jurisdiction over endorsement of the Code has the responsibility of making Code interpretations (see NEC, Articles 90-4).

5-3 When are rules in the NEC mandatory and when are they advisory?
When the word "shall" is used, the rules are mandatory, and when the word "should" is used, they are advisory (see NEC, Section 110-1).

5-4 According to the NEC, What are Voltages?
Throughout this Code, the voltage considered shall be that at which the circuit operates.

5-5 When wire gauge or size is referred to, what wire gauge is used?
The American Wire Gauge (AWG) (see NEC, Article 110-6).

5-6 When referring to conductors, what material is referred to?
To copper, unless otherwise specified. Where other materials are to be used, the wire sizes must be changed accordingly (See the NEC, Section 110-5).

5-7 In what manner must the work be executed?
All electrical equipment must be installed in a neat and workmanlike manner (see NEC, Article 110-12).

BRANCH CIRCUITS AND FEEDERS

5-8 May wooden plugs be used for mounting equipment in masonry, concrete, plaster, etc.?
No (see NEC, Article 110-13).

5-9 How may conductors be spliced or joined together?
They must be spliced or joined together by approved splicing devices or by brazing, welding, or soldering with a fusible metal or alloy (see NEC, Article 110-14).

5-10 When soldering, what precautions must be used?
All joints or splices must be electrically and mechanically secure before soldering and then soldered with a noncorrosive flux (see NEC, Section 110-14).

5-11 How should splices or joints be insulated?
They must be covered with an insulation that is equivalent to the original conductor insulation (see NEC, Section 110-14).

5-12 Can an autotransformer be used on an ungrounded system?
The autotransformer must have an identified grounding conductor that is common to both primary and secondary circuits and is tied into an identified grounding conductor on the system supplying the autotransformer (see NEC, Section 200-9).

An autotransformer may be used to extend or add an individual branch circuit in an existing installation for equipment load without the connection to similar identified grounded conductor when transforming from a nominal 208 volts to a nominal 240 volt supply or similarly from 240 volts to 208 volts.

5-13 On No. 6 or smaller conductors, what means must be used for the identification of the grounding conductors?
Insulated conductors of No. 6 or smaller, when used as grounded conductors, must be white or natural gray. On Type MI cable, the conductors must be distinctively marked at the terminal during installation (see NEC, Article 200-6).

5-14 How should conductors larger than No. 6 be marked to indicate the grounded wire?
By use of white or natural gray colored insulation or by identifying with a distinctive white marking at the terminals during installation (see NEC, Article 200-6).

Branch Circuits and Feeders

5-15 How is the high-leg conductor of a 4-wire Delta identified?

Where, on a 4-wire Delta-connected secondary, the midpoint of one phase is grounded to supply lighting and similar loads, that phase conductor having the higher voltage to ground shall be orange in color or be indicated by tagging or other effective means at the point where a connection is to be made if the neutral conductor is present. (See Section 215-8 of the NEC.)

5-16 On a grounded system, which wire must be connected to the screw shell of a lampholder?

The identified conductor (see NEC, Article 200-10(c)).

5-17 What will determine the classification of branch circuits?

The maximum permitted setting or rating of the overcurrent protection device in the circuit (see NEC, Article 210-3).

5-18 What color coding is required on multiwire branch circuits?

The grounded conductor of a branch circuit shall be identified by a continuous white or natural gray color. Where conductors of different systems are installed in the same raceway, box, auxiliary gutter, or to other types of enclosures, one system neutral, if required, shall have an outer covering of white or natural gray. Each other system neutral, if required, shall have an outer covering of white with an identifiable colored stripe (not green) running along the insulation or other and different means of identification. Ungrounded conductors of different voltages shall be different color or identified by other means.

5-19 How must a conductor that is only used for grounding purposes be identified?

By the use of a green color, or green with one or more yellow stripes, except if it is bare. (See NEC, Article 210-5).

5-20 Can green colored wire be used for circuit wires?

No. Green is intended for identification of grounding conductors only (see NEC, Article 210-5).

5-21 What voltage is used between conductors that supply lampholders of the screw-shell type, receptacles, and appliances in dwellings?

109

Branch Circuits and Feeders

Generally speaking, a voltage of 150 volts between conductors is considered the maximum. There are, however, some exceptions (see NEC, Article 210-6).

5-22 What are the exceptions to the ruling of 150 volts maximum between conductors?

Permanently connected appliances, portable appliances of more than 1,380 watts, and portable motor-operated appliances of ¼-hp rating or greater (see NEC, Article 210-6).

5-23 How shall you ground the grounding terminal of a grounding type receptacle?

By the use of a grounding conductor of green-covered wire, green with one or more yellow stripes, or bare conductors. However, the armor of Type AC metal-clad cable, the sheath of MI or ALS cable, or a metallic raceway is acceptable as a grounding means. Section 350-5 does not permit the general use of flexible metal conduit as a grounding means unless it and the connectors are approved for the purpose. (See NEC, Sections 210-7, 250-45, 250-57(a), 250-59, 250-91(b) and 350-5).

5-24 How can you ground the grounding conductor for a grounding-type receptacle on extensions to existing systems?

Run the grounding conductor to a grounded cold water pipe near the equipment (see NEC, Section 210-7(c) Exception).

5-24a What special requirements has been added for 15- and 20-ampere receptacle outlet circuits on construction sites?

Effective January 1, 1974, approved ground-fault circuit-interrupters shall be installed on all 15 and 20 ampere receptacle outlets on 120 volts. (See Section 210-8(b) of the NEC.)

5-25 What is the minimum size for branch-circuit conductors?

They cannot be smaller than No. 8 for ranges of 8¾ kw or more rating and not smaller than No. 14 for other loads. (See NEC, Section 219-19 to 210-19(b).

5-26 What is the requirement concerning all receptacles on 15- and 20-ampere branch circuits?

BRANCH CIRCUITS AND FEEDERS

All receptacles on 15- and 20-ampere branch circuits must be of the grounding type. A single receptacle installed on an individual branch circuit shall have a rating of not less than the rating of the branch circuit. (For complete details, see NEC, Section 210-7(a) and (b)).

5-27 What are the requirements for spacing receptacles in dwelling occupancies?

All receptacles in kitchens, family rooms, dining rooms, breakfast rooms, living rooms, parlors, libraries, dens, sun rooms, recreation rooms, and bedrooms must be installed so that no point along the wall space, measured horizontally, is more than six feet from a receptacle. This includes wall space that is two feet or wider and any space occupied by sliding panels on exterior walls. At least one outlet shall be installed for the laundry. (See NEC, Section 210-5(b). The wall space afforded by fitted room dividers, such as free-standing bar-type counters, shall be included in the 6-foot measurement.

In kitchen and dining areas a receptacle outlet shall be installed at each counter space wider than 12 inches. Counter-top spaces separated by range-tops, refrigerators or sinks shall be considered as separate counter top space. Receptacles rendered inaccessible by the installation of stationary appliances will not be considered as the required outlets.

At least one wall receptacle outlet shall be installed in the bathroom adjacent to the basin location.

5-27a In residential occupancies, will ground-fault circuit interrupters be required?

Yes, for all 120 volt, 15 and 20 ampere receptacle outlets installed out-of-doors for residential occupancies and also for receptacle outlets in bathrooms. (See the NEC Section 210-8(a).)

5-28 A bedroom has one wall that contains a closet 8 feet in length, with sliding doors, and a wall space of 2½ feet; the bedroom door opens into the room and back against this wall space. Where are receptacles required on this wall?

One receptacle is required in the 2½ foot space (see Fig. 1). (See the NEC, Section 210-25(b).)

BRANCH CIRCUITS AND FEEDERS

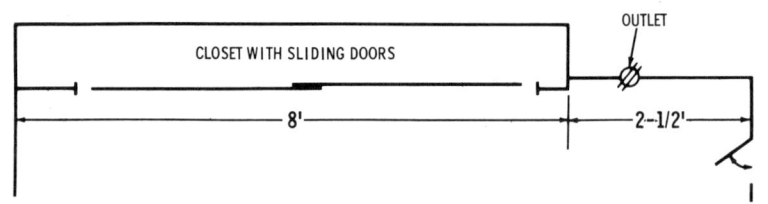

Fig. 1. Receptacle requirements on a wall containing a closet with sliding doors.

5-29 Sketch a typical living room with sliding panels on the outside wall; locate receptacles, and show the room dimensions. (See Fig. 2.)

5-30 A fixed appliance is located on a 15- or 20-amp branch circuit on which there is a lighting fixture. What is the maximum rating that would be permitted on the fixed appliance?

50% of the branch circuit rating (see NEC, Section 210-23(a)).

5-31 A portable appliance is used on a 15- or 20-amp branch circuit. What is the maximum rating permitted for the portable appliance?

80% of the branch circuit rating (see NEC, Section 210-23(a)).

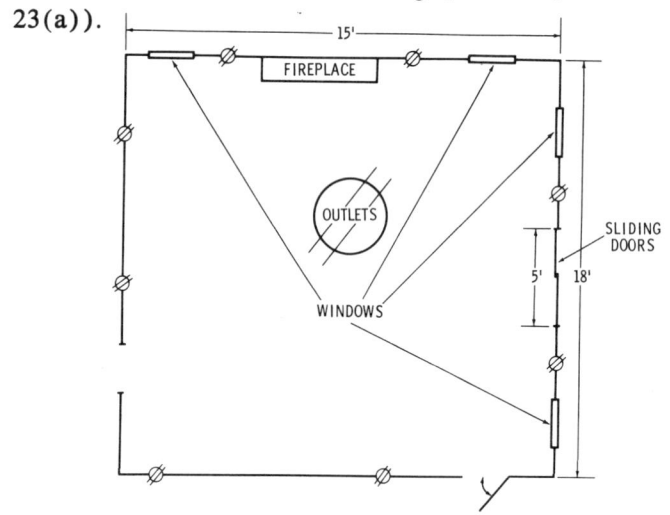

Fig. 2. Receptacle requirements in a living room that has sliding doors on the outside wall.

112

BRANCH CIRCUITS AND FEEDERS

5-32 What is the smallest size wire permissible for a feeder circuit?

Number 10 wire (see NEC, Section 215-2(a)).

5-33 What are the permissible voltage drops allowable on feeders and branch circuits?

On feeders, not more than 3% for power, heating, and lighting loads or combinations thereof. A total maximum voltage drop not to exceed 5% for conductors and for combinations of feeders and branch circuits. (See NEC, Section 215-2(c) fine-print note.)

Table 1. General Lighting Loads by Occupancies

Type of Occupancy	Unit Load per Sq. Ft. (watts)
Armories and Auditoriums	1
Banks	5
Barber Shops and Beauty Shops	3
Churches	1
Clubs	2
Court Rooms	2
Dwellings (other than hotels)	3
Garages—Commercial (storage)	½
Hospitals	2
Hotels, including apartment houses without provisions for cooking by tenants	2
Industrial Commercial (Loft) Buildings	2
Lodge Rooms	1½
Office Buildings	5
Restaurants	2
Schools	3
Stores	3
Warehouse Storage	¼

In any of the above occupancies except single-family dwellings and individual apartments of multi-family dwellings:
   Assembly Halls and Auditoriums
   Halls, Corridors, Closets
   Storage Spaces
	
	1
	½
	¼

Branch Circuits and Feeders

5-34 What is the basis for figuring the general lighting loads in occupancies?

They are figured on a watts-per-square-foot basis, using Table 1 as a reference. The table below can be used to determine the unit load per square foot in watts for the types of occupancies listed.

5-35 What measurements are used to determine the number of watts per square foot?

The outside dimensions of the building and the number of floors, not including open porches or garage (see NEC, Section 220-2(b)).

5-36 Are unfinished basements of dwellings used in figuring watts per square foot?

Yes, if adaptable for future use; if these spaces are not adaptable, then they are not used (see NEC, Article 220-2(b)).

5-37 What loads are used in figuring outlets for other than general illumination?

Outlets supplying specific loads and appliances must use the ampere rating of the appliance. Outlets supplying heavy-duty lampholders must use 600 volt-amperes; for other outlets, they must use 180 volt amperes (see NEC, Section 220-2(c)).

5-38 What load must be figured for show-window lighting?

Not less than 200 watts for each linear foot measured horizontally along the base (see NEC, Section 220-12).

5-39 What are the receptacle requirements in dwelling occupancies for kitchen, family room, laundry, pantry, dining room, and breakfast room?

There shall be a minimum of 2-20 ampere small appliance circuits for the kitchen, family room, pantry, dining room, and breakfast room. Also, a minimum of 1-20 ampere circuit for the laundry. (See NEC, Section 220-3 (b and c).)

5-40 Are demand factors permitted in determining feeder loads?

Yes; they are given in Table 2. Also, see NEC, Section 220-11.

5-41 What size feeders must be installed in dwelling occupancies?

The computed load of a feeder must not be less than the sum

Branch Circuits and Feeders

of all branch circuit loads supplied by the feeder. The demand factors may be used in the calculation of feeder sizes.

5-42 When figuring the neutral load to electric ranges, what is the maximum unbalanced load considered to be?

The maximum unbalanced load for electric ranges is considered to be 70% of the load on the ungrounded conductors (see NEC, Section 220-22).

5-43 How is the neutral load on a 5-wire 2-phase system determined?

It is figured at 140% of the load on the ungrounded conductors (see NEC, Section 220-22).

Table 2. Calculation of Feeder Loads by Occupancies

Type of Occupancy	Portion of Lighting Load to Which Demand Factor Applies (wattage)	Feeder Demand Factor
Dwellings—other than hotels	First 3,000 or less at	100%
	Next 3,001 to 120,000 at	35%
	Remainder over 120,000 at	25%
*Hospitals	First 50,000 or less at	40%
	Remainder over 50,000 at	20%
*Hotels—including apartment houses without provision for cooking by tenants	First 20,000 or less at	50%
	Next 20,001 to 100,000 at	40%
	Remainder over 100,000 at	30%
Warehouses (storage)	First 12,500 or less at	100%
	Remainder over 12,500 at	50%
All Others	Total Wattage	100%

*The demand factors of this table do not apply to the computed load of subfeeders to areas in hospitals and hotels where entire lighting is likely to be used at one time, such as in operating rooms, ballrooms, or dining rooms.

5-44 How may the neutral-feeder load on a 3-wire DC or single-phase 3-wire AC system be determined?

The 70% demand factor may be used on range loads, and a further demand factor of 70% may be used on that portion of the unbalanced load in excess of 200 amperes (see NEC, Section 220-22).

5-45 How do you calculate the unbalanced load on a 4-wire 3-phase system?

BRANCH CIRCUITS AND FEEDERS

The 70% demand factor may be used on range loads, and a further demand factor of 70% may be used on that portion of the unbalanced load in excess of 200 amperes (see NEC, Section 220-22).

5-46 Can you make a reduction on the neutral feeder load where discharge lighting is involved?

No reduction can be made on the neutral capacity for that portion of the load which consists of electric discharge lighting. The load on the neutral feeder for discharge lighting must be taken at 100% of the ungrounded conductors (see NEC, Section 220-22).

5-47 Why do discharge lighting loads require no reduction in neutral feeder capacity?

Because of the effect of the third harmonic on the current value in the neutral feeder.

Table 3. Demand Loads for Household Electric Ranges, Wall-Mounted Ovens, Counter-Mounted Cooking Units, and Other Household Cooking Appliances Over 1¾ kw Rating

(Column A to be used in all cases except as otherwise permitted in Note 4 below.)

Number of Appliances	Maximum Demand (See Notes) Column A (Not over 12 kw rating)	Demand Factor (See Note 4) Column B (Less than 3½ kw rating)	Column C (3½ kw to 8¾ kw rating)
1	8 kw	80%	80%
2	11 kw	75%	65%
3	14 kw	70%	55%
4	17 kw	66%	50%
5	20 kw	62%	45%
6	21 kw	59%	43%
7	22 kw	56%	40%
8	23 kw	53%	36%
9	24 kw	51%	35%
10	25 kw	49%	34%
11	26 kw	47%	32%
12	27 kw	45%	32%
13	28 kw	43%	32%
14	29 kw	41%	32%
15	30 kw	40%	32%

Branch Circuits and Feeders

Table 3. (continued)

Number of Appliances	Maximum Demand (See Notes) Column A (Not over 12 kw rating)	Demand Factor (See Note 4) Column B (Less than 3½ kw rating)	Column C (3½ kw to 8¾ kw rating)
16	31 kw	39%	28%
17	32 kw	38%	28%
18	33 kw	37%	28%
19	34 kw	36%	28%
20	35 kw	35%	28%
21	36 kw	34%	26%
22	37 kw	33%	26%
23	38 kw	32%	26%
24	39 kw	31%	26%
25	40 kw	30%	26%
26-30	15 kw plus 1 kw for each range	30%	24%
31-40		30%	22%
41-50	25 kw plus ¾ kw for each range	30%	20%
51-60		30%	18%
61 & over		30%	16%

Note 1. Over 12 kw to 27 kw ranges all of same kw rating. For ranges, individually rated more than 12 kw but not more than 27 kw, the maximum demand in Column A must be increased 5% for each additional kw of rating or major fraction thereof by which the rating of individual ranges exceeds 12 kw.

Note 2. Over 12 kw to 27 kw ranges of unequal rating. For ranges individually rated more than 12 kw and of different ratings but none exceeding 27 kw, an average value of rating must be calculated by adding together the ratings of all ranges to obtain the total connected load (using 12 kw for any range rated less than 12 kw) and dividing by the total number of ranges; then, the maximum demand in Column A must be increased 5% for each kw or major fraction thereof by which this average value exceeds 12 kw.

Note 3. This table does not apply to commercial ranges. The branch-circuit load for a commercial range must be the nameplate rating of the range.

Note 4. Over 1¾ kw to 8¾ kw. In lieu of the method provided in Column A, loads rated more than 1¾ kw but less than 8¾ kw may be considered as the sum of the nameplate ratings of all the loads multiplied by the demand factors specified in Column B or C for the given number of loads.

5-48 Are demand factors applicable to electric ranges?

Yes. Table 3 may be used in determining demand factors for electric ranges. (See NEC, Table 220-19).

BRANCH CIRCUITS AND FEEDERS

5-49 Can demand factors be used on electric clothes dryer loads in the same manner as they are used on electric ranges?

Yes, see Table 4. (See NEC, Table 220-18).

5-49a Are feeder demand factors permitted for commercial ranges and other commercial kitchen equipment?

Yes, see table 5. (See NEC, Table 220-20).

5-50 There are 2,500 square feet of floor area in a house which contains an electric range rated at 12 kw and an electric dryer rated at 4,500 watts. Calculate: (a) the general lighting load; (b) the minimum number and sizes of branch circuits required; (c) the minimum size of feeders (service conductors) required.

(a) 2,500 sq. ft at 3 watts per sq. ft. equals 7,500 watts.

(b) 7,500 watts divided by 115 volts equals 65 amperes; this would require a minimum of five 15-amp circuits with a minimum of No. 14 wire, or a minimum of four 20-amp circuits with a minimum of No. 12 wire.

Table 4. Demand Factors for Household Electric Clothes Dryers

Number of Dryers	Demand Factor (per cent)
1	100
2	100
3	100
4	100
5	80
6	70
7	65
8	60
9	55
10	50
11-13	45
14-19	40
20-24	35
25-29	32.5
30-34	30
35-39	27.5
40 up	25

BRANCH CIRCUITS AND FEEDERS

Small appliance load:

A minimum of two small-appliance circuits (see NEC, Article 220-3b) of 1,500 watts each; these will require 20-amp 2-wire circuits with No. 12 wire.

A 12-kw range has a demand of 8 kw, according to Table 3. 8,000 watts divided by 230 volts equals 35 amps; therefore, a minimum of No. 8 wire with a 40-amp service would be required, although good practice would indicate the use of No. 6 wire with a 50-amp circuit breaker.

The dryer; 4,500 watts divided by 230 volts equals 20 amps; therefore, you would use No. 12 wire with a 20-amp circuit; however, good practice, with higher wattages coming all the time in dryers, would suggest No. 8 wire with a 40-amp circuit breaker.

Laundry circuit, 1,500 watts.

(c) Minimum size of service.

General lighting	7,500 watts
Small appliance load	3,000 watts
Laundry circuit	1,500 watts
	12,000 watts
3,000 watts @ 100%	3,000 watts
9,000 watts @ 35%	3,150 watts
Net computed (without range and dryer)	6,150 watts
Range load	8,000 watts
Dryer load	4,500 watts
	12,500 watts
Net computed (with range and dryer)	18,650 watts

Loads of over 10 kw must have a minimum of 100-ampere service (see NEC, Article 250-94); therefore, the minimum service will be 100 amp. We can use No. 1 TW wire or No. 3 RH wire, with a 100-ampere main disconnect. We must use a minimum of No. 6 service entrance ground wire with No. 1 entrance wire or a minimum of No. 8 service entrance ground wire with No. 3 entrance wire.

5-51 A store building is to be wired for general illumination and show-window illumination. The store is 40 feet by 75 feet, with 30 linear feet of show window. With a

Branch Circuits and Feeders

density of illumination of 3 watts per square foot for the store and 200 watts per liner foot for the show-window, calculate: (a) the general store load; (b) the minimum number of branch circuits required and sizes of wire; (c) the minimum size of feeders (or service conductors) required.

(a) General lighting load.

3,000 sq. ft. at 3 watts per sq. ft. equals 9,000 watts. However, this load will be required most of the time, so we must multiply 1.25 times 9,000, or 11,250 watts.

Show window:

30 linear feet at 200 watts per foot equals 6,000 watts. Therefore, the general store load is 11,250 watts plus 6,000 watts equals 17,250 watts.

(b) Minimum number of branch circuits and wire sizes.

11,250 watts divided by 230 volts equals 49 amps. For three-wire service, this current will require four 15-amp circuits using No. 14 wire (minimum) or three 20-amp circuits using No. 12 wire (minimum).

Show window:

6,000 watts divided by 230 volts equals 26 amps. Therefore, two 15-amp circuits with No. 14 wire or two 20-amp circuits with No. 12 wire will be required.

(c) Minimum size feeders (service conductors) required.

The ampere load for three-wire service would be 49 amps plus 26 amps equals 75 amps. Therefore, a 100-ampere service would be used with No. 1 TW wire in 1½-inch conduit or No. 3 RH wire in 1¼-inch conduit.

If the service is 115 volts instead of 115/230 volts (which is highly improbable, because the utility serving would require the 115/230 volts), then we would have to use 115 volts when finding the current. These currents would then be doubled; entrance-wire capacities would double as well as the main disconnect on the service entrance.

5-52 Determine the general lighting and appliance load requirements for a multifamily dwelling (apartment house). There are 40 apartments, each with a total of 800 square feet. There are two banks of meters of 20 each,

BRANCH CIRCUITS AND FEEDERS

and individual subfeeders to each apartment. Twenty apartments have electric ranges; these apartments (with ranges) are evenly divided, 10 on each meter bank; the ranges are 9 kw each. The service is 115/230 volts. Make complete calculations of what will be required, from the service entrance on.

Computed load for each apartment (see NEC, Article 220).
    General lighting load:
        800 sq. ft. @ 3 watts per sq. ft. .... 2,400 watts
    Small appliance load .... 3,000 watts
    Electric range .... 3,000 watts

Minimum number of branch circuits required for each apartment (see NEC, Article 220-3).
    General lighting load:
        2,400 watts divided by 115 volts equals 21 amps. This current will require two 15-amp circuits using No. 14 wire or two 20-amp circuits using No. 12 wire.
    Small appliance load:
        Two 20-amp circuits using No. 12 wire (see NEC, Article 220-3b).
    Range circuit:
        8,000 watts divided by 230 equals 34 amps. A circuit of two No. 8 wires or one No. 10 wire is required (see NEC, Article 210-19c).

Minimum size subfeeder required for each apartment (see NEC, Article 215-2).
    Computed load:

        General lighting load .... 2,400 watts
        Small appliance load (two 20-amp circuits) . 3,000 watts
        Total computed load (without ranges) .... 5,400 watts

    Application of demand factor:
        3,000 watts @ 100% .... 3,000 watts
        2,400 watts @ 35% ....   840 watts
        Net computed load (without ranges) ..... 3,840 watts
        Range load .... 8,000 watts
        Net computed load (with ranges) .... 11,840 watts

    For 115/230 volt, 3-wire system (without ranges):

BRANCH CIRCUITS AND FEEDERS

Net computed load
3,840 watts divided by 230 volts equals 16.7 amps.
Minimum feeder size would be No. 10 wire with a two-pole, 30-amp circuit breaker.

For 115/230 volt, 3-wire system (with ranges):
Net computed load
11,840 watts divided by 230 volts equals 51.5 amps.
Minimum feeder size would be No. 6 wire with two 60-amp fuses or No. 4 wire with a two-pole, 70-amp circuit breaker.

Neutral subfeeder:

Lighting and small-appliance load .... 3,840 watts
Range load (8,000 watts @ 70%) (see NEC, Section 220-22) .... 5,600 watts
Net computed load (neutral) .... 9,440 watts

9,440 watts divided by 230 volts equals 41 amps. Size of neutral subfeeder would be No. 6 wire. There would also be two main disconnects needed ahead of the meters.

Minimum size feeders required from service equipment to meter bank (for 20 apartments—10 with ranges).

Total computed load:
Lighting and small-appliance load—20 times 5,400 watts equals 108,000 watts

Application of demand factor:

3,000 watts @ 100% .... 3,000 watts
105,000 watts @ 35% .... 36,750 watts
Net computed lighting and small-appliance load .... 39,750 watts
Range load (10 ranges, less than 12 kw) .. 25,000 watts
Net computed load (with ranges) .... 64,750 watts

For 115/230 volt, 3-wire system:
Net computed load, 64,750 divided by 230 equals 282 amps. Size of each ungrounded feeder to each meter bank would be 500,000 CM.

Neutral feeder:
Lighting and small-appliance load .... 39,750 watts
Range load (25,000 watts @ 70%) ...... 17,500 watts
Computed load (neutral) .... 57,250 watts

Branch Circuits and Feeders

57,250 watts divided by 230 volts equals 249 amps.

Further demand factor:

200 amps @ 100%	200 amps
49 amps @ 70%	34 amps
Net computed load (neutral)	234 amps

Neutral feeder to each meter bank would be 300,000 CM. Minimum size main feeder (service conductors) required (for 40 apartments—20 with ranges).

Total computed load:

Lighting and small-appliance load—40 times 5,400 watts equals 216,000 watts

Application of demand factor:

3,000 watts @ 100%	3,000 watts
117,000 watts @ 35%	40,950 watts
96,000 watts @ 25%	24,000 watts
Net computed lighting and small-appliance load	67,950 watts
Range load (20 ranges, less than 12 kw)	35,000 watts
Net computed load	102,950 watts

For 115/230 volt, 3-wire system:

Net computed load—102,950 watts divided by 230 volts equals 448 amps.

Size of each ungrounded main feeder would be 1,000,000 CM.

Neutral feeder:

Lighting and small-appliance load	67,950 watts
Range load (35,000 watts @ 70%)	24,500 watts
Computed load (neutral)	92,450 watts

92,450 watts divided by 230 volts equals 402 amps.

Further demand factor:

200 amps @ 100%	200 amps
202 amps @ 70%	141 amps
Net computed load (neutral)	341 amps

Size of neutral main feeder would be 600,000 CM.

5-53 There is to be a current of 100 amperes per phase on a

BRANCH CIRCUITS AND FEEDERS

4-wire 120/208 volt Wye system. What size neutral would we need? (The phase wires are No. 2 RH.)

We would use No. 2 RH wire (see NEC, Section 220-22).

5-54 Is there a demand factor for feeders and service entrance conductors for multi-family dwellings?

Yes, see Table 6. (See NEC, Table 220-32.)

Table 5. Feeder Demand Factors for Commercial Electric Cooking Equipment, Including Dishwasher Booster Heater, Water Heaters, and Other Kitchen Equipment

Number of Units	Demand Factors (per cent)
1	100
2	100
3	90
4	80
5	70
6 & Over	65

Table 6. Demand Factors for Feeders and Service Entrance Conductors for Multifamily Dwellings

Number of Dwelling Units	Demand Factor (per cent)
3- 5	45
6- 7	44
8-10	43
11	42
12-13	41
14-15	40
16-17	39
18-20	38
21	37
22-23	36
24-25	35
26-27	34
28-30	33
31	32
32-33	31
34-36	30
37-38	29
39-42	28
43-45	27
46-50	26
51-55	25
56-61	24
62 & Over	23

CHAPTER 6

Transformer Principles and Connections

6-1 What is induction?

The process by which one conductor produces, or induces, a voltage in another conductor, even though there is no mechanical coupling between the two conductors.

6-2 What factors affect the amount of induced electromotive force (emf) in a transformer?

The strength of the magnetic field, the speed at which the conductors are cut by the magnetic field, and the number of turns of wire being cut by the magnetic field.

6-3 What is inductance?

The property of a coil in a circuit to oppose any change of existing current flow.

6-4 What is self-inductance?

The inducing of an emf within the circuit itself, caused by any change of current within that circuit. This induced emf is always in a direction opposite to the applied emf, thus causing opposition to any change in current within the circuit itself.

6-5 What is mutual inductance?

The linkage of flux between two coils, caused by the current within one coil.

6-6 Draw a diagram of two coils, such as the coils of a transformer winding, and indicate the self-inductance and the mutual inductance.

Self-inductance is produced within the primary coil and mutual inductance exists between the two transformer coils, as shown in Fig. 1.

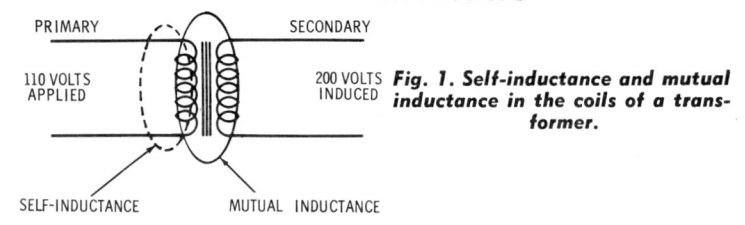

Transformer Principles and Connections

Fig. 1. Self-inductance and mutual inductance in the coils of a transformer.

6-7 Name several methods by which an emf may be generated.

By conductors being cut by a magnetic field (generators), by chemical action (batteries), by heat (thermocouples), by crystal vibration (phono cartridges), and by friction.

6-8 What is direct current (DC)?

Current that flows in one direction only.

6-9 What is alternating current (AC)?

Current that continually reverses its direction of flow.

6-10 What is pulsating direct current?

A unidirectional current that changes its value at regular or irregular intervals.

6-11 What is a cycle?

One complete alternation, or reversal, of alternating current. The wave rises to maximum in one direction, from zero, falls back to zero, then rises to maximum in the opposite direction, and finally falls back to zero again.

6-12 What must surround a conductor when current flows through it?

A magnetic field.

6-13 What is the phase relation between the three phases of a three-phase circuit?

They are 120 electrical degrees apart.

6-14 Draw sine waves for three-phase voltage; show polarity, time, and phase angle (in degrees). (See Fig. 2.)

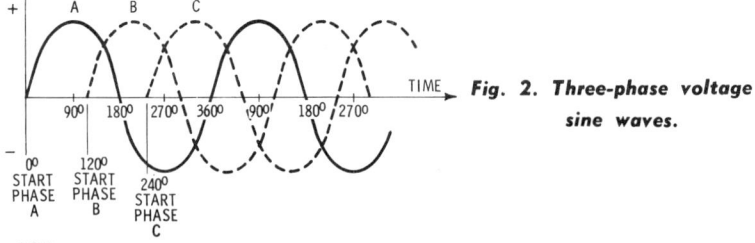

Fig. 2. Three-phase voltage sine waves.

126

Transformer Principles and Connections

6-15 What is the phase relation between phases in a two-phase circuit?

They are 90 electrical degrees apart.

6-16 What is a transformer?

A device that transforms electrical energy from one or more circuits to one or more other circuits at the same frequency but usually at a different voltage and current. It consists of a core of soft-iron laminations, surrounded by coils of copper-insulated wire.

6-17 There are two basic types of transformers. What are they?

The isolation type—the two windings are physically isolated and electrically insulated from each other; and the autotransformer type—there is only one coil with a tap or taps taken off of it to secure other voltages (the primary is part of the secondary and the secondary is part of the primary).

6-18 What is an oil-immersed transformer?

The core and coils are immersed in a high-grade mineral oil, which has high dielectric qualities.

6-19 Why is oil used in a transformer?

To increase the dielectric strength of the insulation, to keep down the possibility of arcing between coils, and to dissipate heat to the outer case so that the transformer may carry heavier loads without excessive overheating.

6-20 What is an air-core transformer?

A transformer that does not contain oil or other dielectric compositions but is insulated entirely by the winding insulations and air.

6-21 What are eddy currents?

Circulating currents induced in conducting materials by varying magnetic fields.

6-22 Are eddy currents objectionable?

Yes; they represent a loss in energy and also cause overheating.

6-23 What means can be taken to keep eddy currents at a minimum?

The iron used in the core of an alternating-current transformer is laminated, or built up of thin sheets or strips of iron, so that eddy currents will only circulate in limited areas.

127

Transformer Principles and Connections

6-24 What is hysteresis?

When iron is subjected to a varying magnetic field, the magnetism lags the magnetizing force, due to the fact that iron has reluctance, or resistance, to changes in magnetic densities.

6-25 Is hysteresis objectionable?

Yes, it is a loss and affects the efficiency of transformers.

6-26 Are transformers normally considered to be efficient devices?

Yes; they have one of the highest efficiencies of any electrical device.

6-27 What factors constitute the major losses produced in transformers?

Power loss of the copper (I^2R losses), eddy currents, and hysteresis losses.

6-28 Is there a definite relationship between the number of turns and voltages in transformers?

Yes; the voltage varies in exact proportion to the number of turns connected in series in each winding.

6-29 Give an illustration of the relationship between the voltages and the turns ratio in a transformer.

If the high-voltage winding of a transformer has 1,000 turns, and a potential of 2,400 volts is applied across it, the low-voltage winding of 100 turns will have 240 volts induced across it; this is illustrated in Fig. 3.

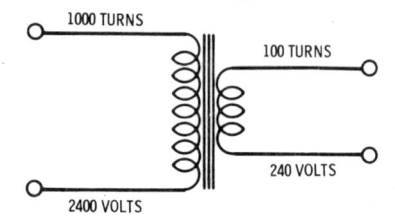

Fig. 3. Relationship between voltages and the turns ratio in a transformer.

6-30 What is the difference between the primary and the secondary of a transformer?

The primary of the transformer is the input side of the transformer and the secondary is the output side of the transformer. On a step-down transformer, the high-voltage side is the primary and the low-voltage side is the secondary; on a step-up transformer, the opposite is true (see Fig. 4).

Transformer Principles and Connections

Fig. 4. The primary and secondary windings of a step-down and a step-up transformer.

6-31 Ordinarily, what is the phase relationship between the primary and secondary voltages of a transformer?

They are 180° out of phase.

6-32 Is it possible to have the primary and secondary of a transformer in phase?

Yes, by changing the connections on one side of the transformer.

6-33 How are the leads of a transformer marked, according to ANSI (American National Standards Institute)?

The high side of the transformer is marked H_1, H_2, etc. The low side of the transformer is marked X_1, X_2, etc.

6-34 What is the purpose of the markings on transformer leads?

They are there for standardization, so that transformer polarities may be recognizable for any type of use.

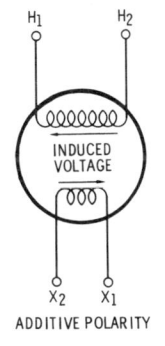

Fig. 5. A transformer with additive polarity.

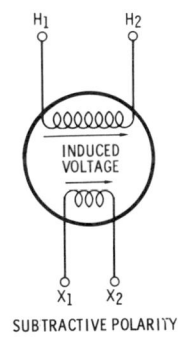

Fig. 6. A transformer with subtractive polarity.

6-35 Draw a diagram of a transformer with additive polarity, using ANSI markings. (See Fig. 5.)

129

Transformer Principles and Connections

6-36 Draw a diagram of a transformer with subtractive polarity, using ANSI markings. (See Fig. 6.)

6-37 If a transformer is not marked, how could you test it for polarity?

Connect the transformer as shown in Fig. 7. If it has subtractive polarity, V will be less than the voltage of the power source; if it has additive polarity, V will be greater than the voltage of the power source.

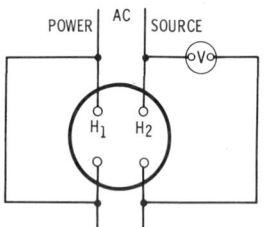

Fig. 7. Testing a transformer for polarity.

6-38 What is a split-coil transformer?

A transformer that has the coils on the low or high side in separate windings, so that they can be connected in series or parallel for higher or lower voltages, as desired.

6-39 Draw a diagram of a split-coil transformer with the low side having split coils for dual voltages; draw an additive-polarity transformer, and mark the terminals with ANSI markings. Show the voltages that you use. (See Fig. 8.)

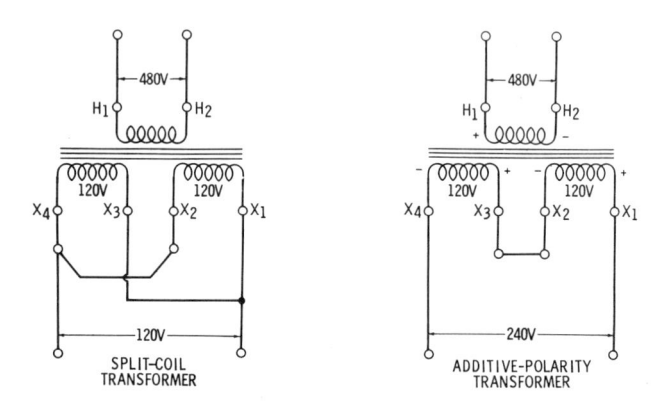

Fig. 8. A split-coil transformer and an additive-polarity transformer.

Transformer Principles and Connections

6-40 Draw a diagram of an autotransformer. (See Fig. 9.)

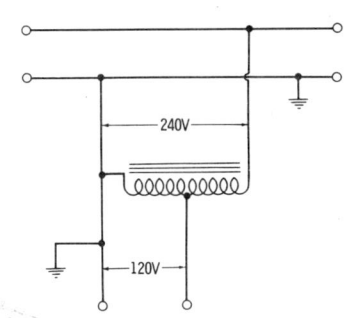

Fig. 9. An autotransformer.

6-41 Where may autotransformers be used?

(a) Where the system being supplied contains an identified grounded conductor that is solidly connected to a similar identified grounded conductor of the system supplying the autotransformer (see NEC, Section 210-9); (b) where an autotransformer is used for starting or controlling an induction motor (see NEC, Article 430-82b); (c) where an autotransformer is used as a dimmer, such as in theaters (see NEC, Article 520-25c); (d) as part of a ballast for supplying lighting units (see NEC, Section 410-78). For voltage bucking and boosting see NEC, Section 210-9 Exception.

6-42 What is the relationship between the current and voltage in the high side of a transformer and the current and voltage in the low side of a transformer? Draw a diagram showing this relationship.

The current in one side of a transformer is inversely proportional to the current in the other side, with respect to the turns ratio, while the voltage across one side of a transformer is directly proportional to the voltage across the other side, with respect to the turns ratio; these are illustrated in Fig. 10.

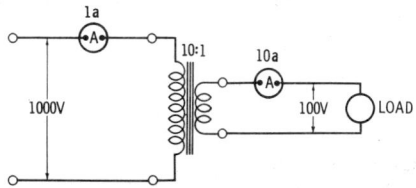

Fig. 10. Current-voltage relationship between the high side and the low side of a transformer.

131

Transformer Principles and Connections

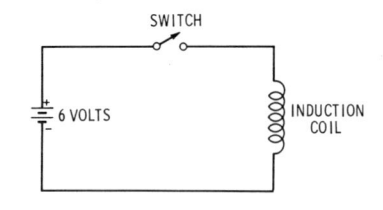

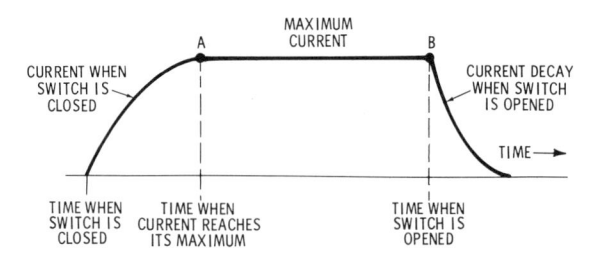

Fig. 11. An induction coil in a DC circuit with the switch being opened and closed.

6-43 When an induction coil is connected in a DC circuit, such as in Fig. 11, what happens when the switch is closed? When the switch is opened?

When the switch is closed, the current slowly rises to a maximum (point A in this example). The retarding of current flow is due to self-inductance. After reaching the maximum at point A, the current will remain constant until the switch is opened (point B). When the switch is opened, the flux around the coil collapses, thereby causing an opposition to the current discharge; however, this discharge-time collapse is extremely short when compared to the charging time. The discharge causes a high voltage to be applied across the switch, which tends to sustain an arc; this voltage often reaches large values. The principles of this type of circuit have many applications, such as ignition coils and flyback transformers.

6-44 Draw a schematic diagram of the high-side windings of three single-phase transformers connected in a Delta arrangement. Show the ANSI markings.

Note that in a Delta arrangement, as shown in Fig. 12, H_1

Transformer Principles and Connections

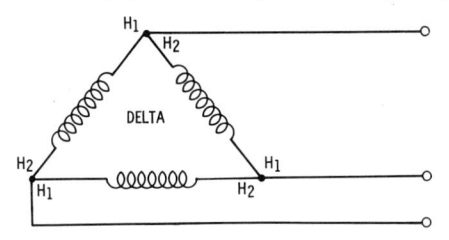

Fig. 12. Three single-phase transformer windings connected in a Delta arrangement.

is connected to H_2 of the next transformer, and so on.

6-45 Draw a schematic diagram of the high-side windings of three single-phase transformers connected in a Wye (y) arrangement. Show the ANSI markings.

Note that in a Wye arrangement, as shown in Fig. 13, all of the H_2's are connected in common, and the H_1's each supply one phase wire.

6-46 If the polarity of one transformer is reversed on a Delta bank of single-phase transformers that are connected for three-phase operation, what would be the result?

Instead of zero voltage on the tie-in point, there would be a voltage that is twice the proper value.

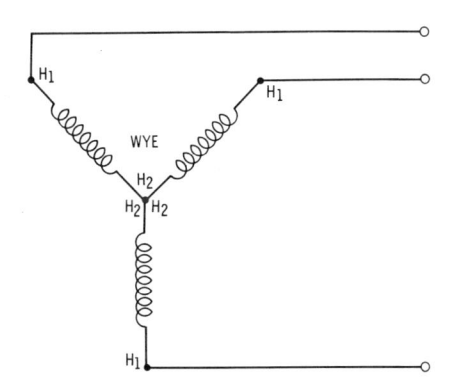

Fig. 13. Three single-phase transformer windings connected in a Wye arrangement.

6-47 Show with diagrams how you would test to be certain

133

TRANSFORMER PRINCIPLES AND CONNECTIONS

that the polarities on a Delta bank of three single-phase transformers are correct. (See Fig. 14.)

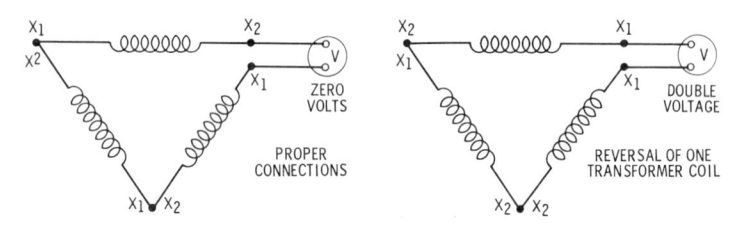

Fig. 14. Polarity tests on a Delta bank of three single-phase transformer windings.

6-48 A bank of three single-phase transformers are connected in a Delta; each transformer delivers 240 volts at 10 amperes. What are the line voltages and line currents?

The line voltages are each equal to 240 volts; however, the line current in each phase would be the current of each transformer times 1.73, or 17.3 amperes.

6-49 Draw a schematic diagram showing all the currents and voltages on a bank of three single-phase transformers that are connected in a Delta arrangement. Assume a voltage across each transformer of 240 volts and a current through each transformer of 10 amperes. (See Fig. 15.)

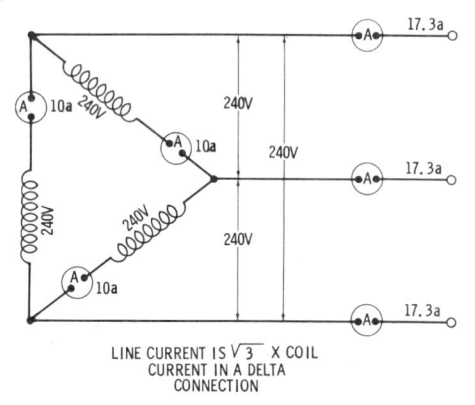

Fig. 15. Current and voltage values on a Delta bank of three single-phase transformer windings. A voltage of 240 volts and a current of 10 amperes are assumed.

TRANSFORMER PRINCIPLES AND CONNECTIONS

6-50 Draw a schematic diagram for a bank of three single-phase transformers that are connected in a Wye. Show all voltages and currents; assume a voltage across each transformer of 120 volts and a current in each winding of 10 amperes. (See Fig. 16.)

6-51 Draw a bank of three single-phase transformers that are connected in a Delta-Delta bank with one side connected to 2,400 volts, three-phase, and with the other side

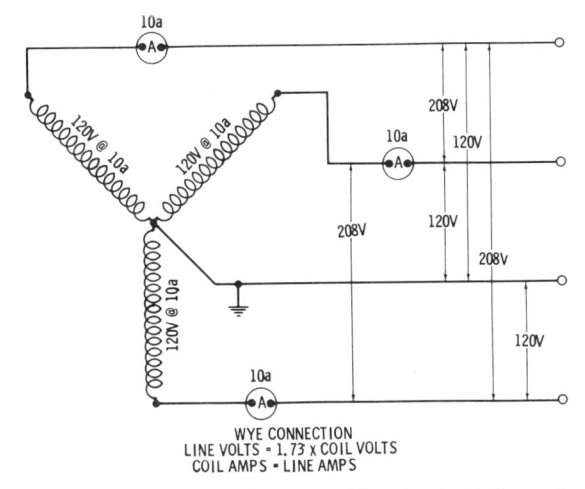

Fig. 16. Current and voltage values on a Wye bank of three single-phase transformer windings. A voltage of 120 volts and a current of 10 amperes are assumed.

delivering 240 volts, three-phase. Show voltages and A.S.A. markings on all transformers. (See Fig. 17.)

6-52 Is it possible to connect two single-phase transformers to secure a three-phase output from a three-phase input?

Yes; they would have to be connected in an open Delta.

6-53 If you have a bank of three single-phase transformers that are connected in a closed Delta arrangement, and one transformer burns up, how would you continue operation on the remaining two transformers?

By merely disconnecting the leads to the disabled transformer.

6-54 When you use a bank of two single-phase transform-

135

TRANSFORMER PRINCIPLES AND CONNECTIONS

ers in an open Delta arrangement, do they supply their full output rating?

No. Each transformer is only capable of supplying 86.6% of its output rating.

6-55 If you have a bank of three single-phase transformers, each with a 10-KVA rating, that are connected in a closed Delta arrangement, you would have a capacity of 30

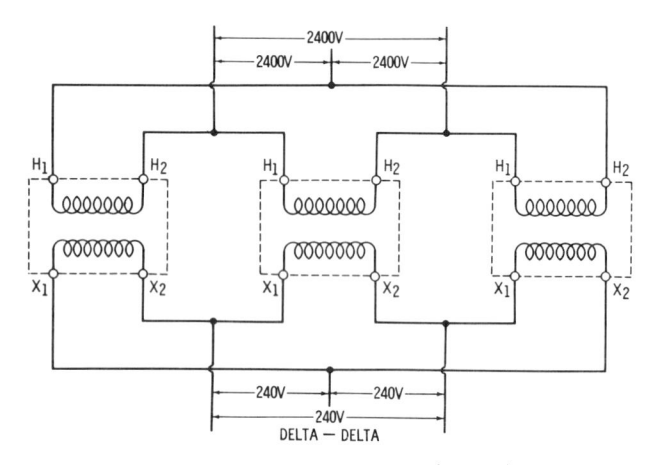

Fig. 17. Three single-phase transformers connected in a Delta-Delta bank; the high side is connected to 2400 volts, 3 phase, and the low side delivers 240 volts, 3 phase.

KVA. If one transformer is taken out of the bank, what would be the output capacity of the remaining 10-KVA transformers?

Each transformer would deliver 8.66 KVA, or you would have a bank capacity of 17.32 KVA.

6-56 Draw a schematic diagram of two transformers that are connected in an open Delta arrangement; show transformer voltages and the three-phase voltages. (See Fig. 18.)

6-57 Draw a schematic diagram of three transformers that are connected in a Delta arrangement on both sides, fed from a 2,400-volt source on the high side, and connected for a 240-volt, three-phase and a 120/240-volt, single-phase output on the low side. Show all voltages. (See Fig. 19.)

Transformer Principles and Connections

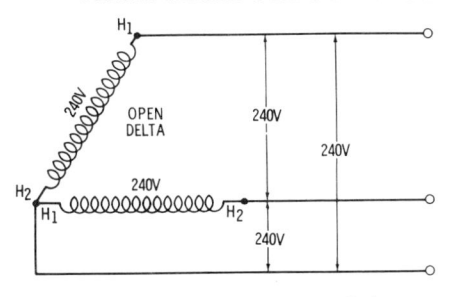

Fig. 18. Two transformers in an open-Delta arrangement.

6-58 Draw a schematic diagram of three single-phase transformers that are connected in a Wye-Wye arrangement. Show the neutral on both high and low sides. (See Fig. 20.)

6-59 Draw a schematic diagram of three single-phase

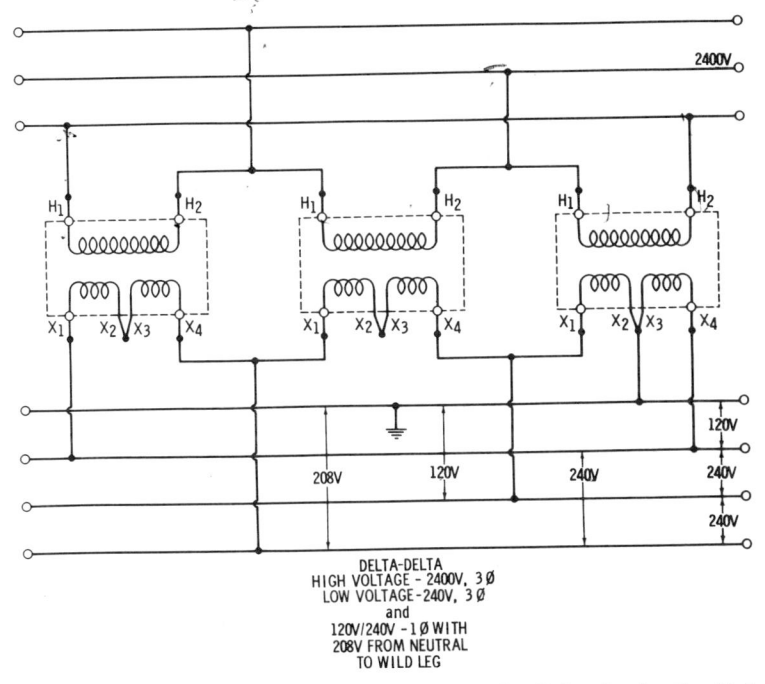

DELTA-DELTA
HIGH VOLTAGE - 2400V, 3 Ø
LOW VOLTAGE - 240V, 3 Ø
and
120V/240V - 1 Ø WITH
208V FROM NEUTRAL
TO WILD LEG

Fig. 19. Three transformers connected in a Delta-Delta bank; the high side is connected to 2400 volts, 3 phase, and the low side delivers 240 volts, 3 phase, and 120/240 volts, 3 phase.

137

transformers that are connected in a Wye arrangement on the high side and a Delta arrangement on the low side. (See Fig. 21.)

6-60 What are instrument transformers?

In the measurement of current, voltage, or kilowatt-hours on systems with high voltage or high current, it is necessary to use a device known as an instrument transformer, which reproduces, in its secondary circuit, the primary current or voltage while preserving the phase relationship to measure or record at lower voltages or lower amperages, and then to use a constant to multiply the readings to obtain the actual values of voltage or current. Current transformers (CT's) are used to measure the current, and potential transformers (PT's) are used to register the potential.

6-61 Describe a potential transformer.

A potential transformer is built like the ordinary isolation transformer, except that extra precautions are taken to assure that the winding ratios are exact. Also, the primary winding is connected in parallel with the circuit to be measured.

6-62 Describe a current transformer.

A current transformer has a primary of a few turns of heavy conductor capable of carrying the total current, and the

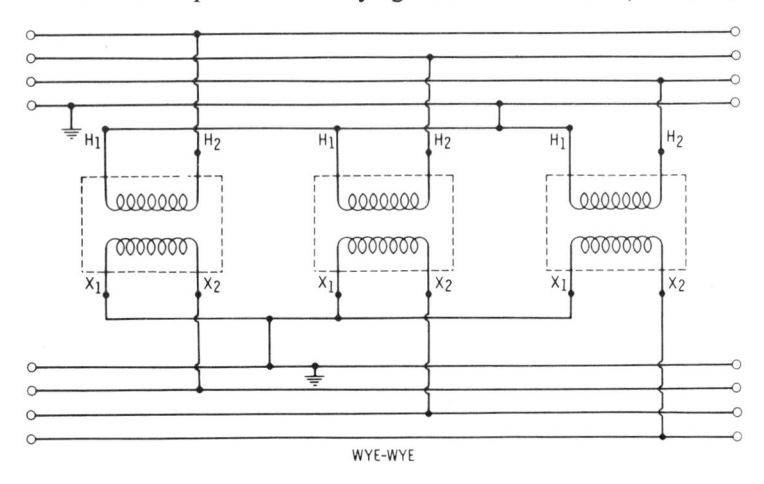

Fig. 20. Three single-phase transformer windings connected in a Wye-Wye arrangement.

TRANSFORMER PRINCIPLES AND CONNECTIONS

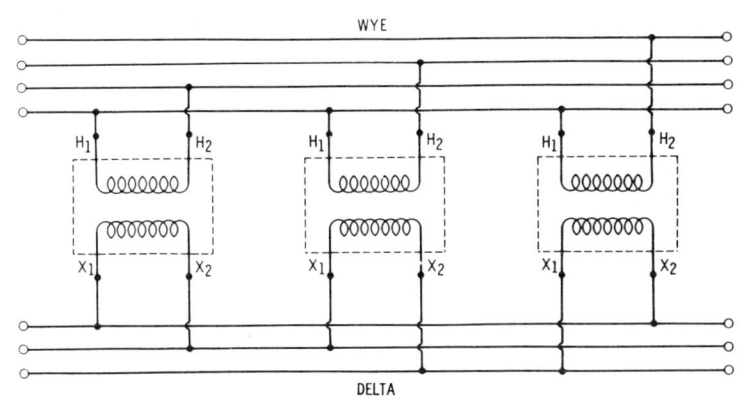

Fig. 21. Three single-phase transformer windings connected in a Wye arrangement on the high side and in a Delta arrangement on the low side.

secondary consists of a number of turns of smaller wire. The primary winding is connected in series with the circuit carrying the current that is to be measured.

6-63 How are current transformers rated?

They are rated at 50 to 5, 100 to 5, etc. The first number is the total current that the transformer is supposed to handle, and the second figure is the current on the secondary when the full-load current is flowing through the primary. For example, a 50-to-5 rating would have a multiplier of 10 ($K = 10$).

6-64 What precautions must be taken when working with current transformers? Why?

The secondary must never be opened when the primary circuit is energized. If it is necessary to disconnect an instrument while the circuit is energized, the secondary must be short circuited. If the secondary is opened while the circuit is energized, the potential on the secondary may reach dangerously high values. By short circuiting the secondary, damage is avoided and the voltage on the secondary is kept within safe limits.

6-65 Draw a schematic diagram of a current transformer.

As shown in Fig. 22, the primary consists of a single conductor; it may be a single conductor or only a few turns.

6-66 What is a booster transformer?

A transformer arrangement that is often used toward the end of a power line in order to raise the voltage back up to its desired value.

Transformer Principles and Connections

Fig. 22. A current transformer.

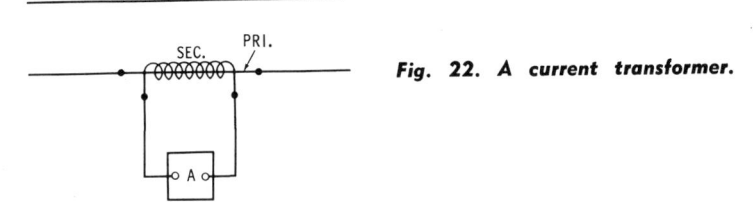

6-67 May an ordinary transformer be used as a booster transformer?
Yes.

6-68 When connecting an ordinary transformer as a booster transformer, what important factors must be considered?
The high side of the transformer must be able to handle the approximate voltage of the line; the low side must have a voltage of approximately the value that you wish to boost the line voltage, and the low side must also have a current capacity that is sufficient to carry the line current.

6-69 What special precaution must be taken when using a booster transformer?
There must be no fusing in the high side, or primary; since the booster transformer is similar to a current transformer, if the fuse should blow, an extremely high voltage could be built up on the secondary side.

6-70 Draw a schematic diagram of an ordinary transformer that is connected as a booster transformer. (See Fig. 23.)

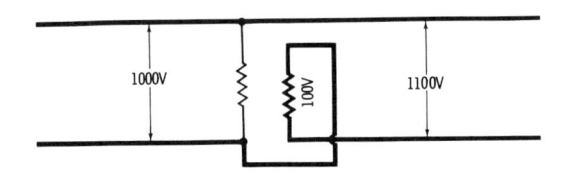

Fig. 23. A voltage-booster transformer.

6-71 What is an induction regulator?
This is a device similar to a booster transformer. It has a primary and a secondary winding which are wound on separate

cores. The primary can be moved in either direction; this is usually done by an electric motor. In turning, the primary bucks or boosts the line voltage, as required. The amount of bucking or boosting is anticipated by the current being drawn by the line.

6-72 Draw a schematic diagram showing how an induction regulator is connected into the line. (See Fig. 24.)

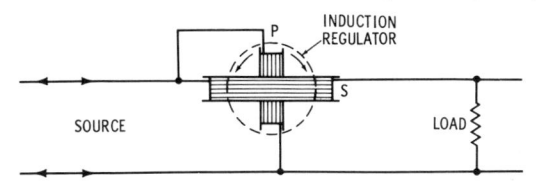

Fig. 24. The connection of an induction regulator to the line.

6-73 What is a three-phase transformer?

A transformer that is the equivalent of three single-phase transformers, which are all wound on one core and enclosed within one common case.

6-74 When connecting transformers in parallel, what factors must be taken into consideration?

Their electrical characteristics, such as voltage ratio, impedance percentage, and voltage regulation.

6-75 If transformers with different electrical characteristics are connected in parallel, what will happen?

They will not distribute the load equally; one transformer will tend to assume more of the load than the other.

CHAPTER 7

Wiring Design and Protection

7-1 What are the minimum requirements for service-drop conductors?

They must be of ample size to carry the load that is required of them, but they must not be smaller than No. 8 copper wire or No. 6 aluminum, except under limited load conditions where they may not be smaller than No. 12 hard-drawn copper (see NEC, Article 230-23).

7-2 Are the conductors from a pole on which the meter or service switch is installed considered as a service drop?

Yes, and they must be installed accordingly (see NEC, Section 230-21).

7-3 What is the minimum clearance for service drops over buildings?

They shall have a minimum clearance of 8 feet. (See NEC, Article 230-24a.)

7-3a What is exception No. 1 to question 7-3?

If the voltage does not exceed 300 volts between conductors and the roof has a slope of not less than 4 inches in 12 inches, the clearance may be a minimum of 3 feet.

7-4 What is the minimum height of point of attachment of service drops?

10 feet, provided the clearance in Section 230-24b is met.

7-5 What is the minimum clearance of service drops over commercial areas, parking lots, agricultural, and other areas subject to truck traffic?

15 feet. (See NEC, Article 230-24b.)

7-6 What is the minimum clearance of service drops over sidewalks?

10 feet (see NEC, Article 230-24b).

143

Wiring Design and Protection

7-7 What is the minimum clearance of service drops over driveways, alleys, and public roads?

18 feet (see NEC, Article 230-24b).

7-8 What is the minimum clearance of service drops over residential driveways?

12 feet (see NEC, Article 230-24b).

7-9 Can a bare neutral conductor be buried in the ground in an underground service?

No; it must be insulated, unless it is in a duct or conduit (see NEC, Article 230-30). There is an exception; bare copper for direct burial where bare copper is judged to be suitable for the soil conditions.

7-10 Where underground service conductors are carried up a pole, up to what minimum height must they be given mechanical protection?

8 feet (see NEC, Section 300-5).

7-11 Is sealing required where underground ducts or conduits enter buildings?

Yes, to prevent the entrance of gases or moisture into the building. Spare or unused conduits and ducts must also be sealed (see NEC, Section 230-48).

7-12 What is the size of the smallest service-entrance conductor normally allowed?

No. 6 wire (see NEC, Article 230-41).

7-13 What is the minimum size service for a single-family dwelling with an initial load of 10 kw or more?

The service must be a minimum of 100 amperes, 3-wire (see NEC, Article 230-41).

7-13a A residence has over six (6) 2-wire branch circuits? What is the minimum size of service required for this residence?

The service entrance conductors shall have an ampacity of not less than 100 amperes, 3-wire. (See NEC, Section 230-41(b)(2).)

7-14 Are splices permitted in service-entrance conductors?

No, they must not be spliced; where they supply a meter, they may be broken as necessary for connecting the meter (see NEC, Section 230-46).

WIRING DESIGN AND PROTECTION

7-15 Are conductors other than service-entrance conductors permitted in the same raceways or cables?

No (see NEC, Section 230-47).

7-17 What must be provided for the disconnection of service conductors from building conductors?

Some means must be provided for disconnecting all of the building conductors from the service-entrance conductors, and this means must be located in a readily accessible point nearest to the point of entrance of the service conductors, on either the outside or inside of the building, whichever is most convenient see NEC, Section 230-72(c)). Also see definition in Article 100 for Service Equipment.

7-18 What may the service-entrance disconnecting means consist of?

It may consist of not more than six switches or six circuit breakers in a common enclosure or in a group of separate enclosures, provided that they are grouped together. Two or three single-pole switches or circuit breakers may be installed on multiwire circuits and counted as one, provided they have handle ties or handles within 1/16 inch proximity to each other, so that not more than six operations of the hand are required to disconnect all circuits (see NEC, Section 230-72(a)).

7-19 What may be connected ahead of the service-entrance disconnecting switch?

Service fuses, meters, high-impedance shunt circuits (such as potential coils of meters), supply conductors for time switches, surge-protection capacitors, instrument transformers, lightning arresters, and circuits for emergency systems (such as fire-pump equipment, etc.) (see NEC, Section 230-82).

7-20 In multiple-occupancy dwellings, is it required that each occupant have access to his disconnecting means?

Yes (see NEC, Section 230-72(d)).

7-21 Is it permissible to install overcurrent-protection devices in the grounded service conductor?

Overcurrent-protection devices may not be installed in the grounded service conductor, unless a circuit breaker is used that opens all conductors of the circuit simultaneously (see NEC, Article 230-90b).

WIRING DESIGN AND PROTECTION

7-22 What is the minimum size of service-entrance conductors allowable?

They must not be smaller than No. 6 wire, unless in cable, where they may be a minimum of No. 8 (see NEC, Section 230-202).

7-24 What is the maximum circuit protection allowed with flexible cords, sizes No. 16 or No. 18, and tinsel cord?

20 amperes (see NEC, Section 240-4).

7-25 What is the maximum overload protection allowed on cords of 10 amperes capacity and 20 amperes capacity?

30-ampere circuit for a 10-ampere cord, and 40-ampere circuits for a 20-ampere cord (see NEC, Section 240-4).

7-26 Give the standard ratings (in amperes) for fuses and circuit breakers?

Standard ratings are: 15, 20, 25, 30, 35, 40, 45, 50, 60, 70, 80, 90, 100, 110, 125, 150, 175, 200, 225, 250, 300, 350, 400, 450, 500, 600, 700, 800, 1000, 1200, 1600, 2000, 2500, 3000, 4000, 5000, 6000. (See NEC, Section 240-6.)

7-27 Where can an overload device be used in a grounded conductor?

No overcurrent device shall be placed in any permanently grounded conductor, except as follows:

Where the overcurrent device simultaneously opens all conductors of the circuit. See Section 240-22 of the NEC.

A fuse shall also be inserted in the grounded conductor when the supply is 3-wire, 3-phase AC, one conductor grounded. See Section 430-36 of the NEC.

7-27a May fuses be arranged in parallel?

No. (See NEC, Section 240-8.)

7-27b May breakers be paralleled?

Only circuit breakers assembled in parallel which are tested and approved as a single unit. (Section 240-14).

7-27c Are the secondary conductors of transformer feeder taps required to have overcurrent protection?

Yes; see Exception No. 8 of Section 240-21 of the NEC. Also note the other 4 exceptions. (See Section 240-3 Exception No. 5 of the NEC.)

Wiring Design and Protection

7-28 In what position must a knife switch with fuses be mounted? Why?

In a vertical position, so that when the switch is opened, the blades will not close by gravity; the line side must be connected so that when the switch is opened, the fuses will be de-energized. (See Section 240-33 of the NEC).

7-29 When overload-device enclosures are mounted in a damp location, what precautions must be taken?

The enclosures must be of an approved type for the location and must be mounted with at least a ¼-inch air space between the wall or supporting surface (see NEC, Article 240-32).

7-30 Where may Edison-base plug fuses be used?

Plug fuses of the Edison-base type shall be used only for replacement in existing installations where there is no evidence of overfusing or tampering.

7-31 What is the maximum voltage rating of plug fuses?

125 volts (see NEC, Section 240-50).

7-32 What precautions must be taken if plug fuses are used on new installations?

Fuses, fuseholders, and adapters must be so designed that other type S fuses may not be used (see NEC, Section 240-52).

7-33 What are the size classifications of S-type fuses and adapters?

Not exceeding 125 volts at 0-15 amperes, 16-20 amperes, and 21-30 amperes (see NEC, Section 240-53(a)).

7-34 What is an S-type fuse?

A dual element fuse with special threads so that fuses larger than what the circuit was designed for may not be used. They are designed to make tampering or bridging difficult.

7-35 Give the classifications of cartridge fuses and fuseholders.

See Table 1.

7-36 Why are systems and circuits grounded?

To limit the excess voltage to ground, which might occur from lightning or exposure to other higher voltage sources (see NEC, Article 250-1).

7-37 Is it necessary to ground one wire on a two-wire DC system with not more than 300 volts between conductors?

Yes, in practically every case, although there are a few exceptions (see NEC, Article 250-3).

WIRING DESIGN AND PROTECTION

Table 1.

Classification.		
(1) 0-600 ampere cartridge fuses and fuseholders are classified with regard to current and voltage as follows:		
Not over 250 volts Amperes	Not over 300 volts Amperes	Not over 600 volts Amperes
0-30	0-15	0-30
31-60	16-20	31-60
61-100	21-30	61-100
101-200	31-60	101-200
201-400		201-400
401-600		401-600
(2) 601-6,000 ampere cartridge fuses and fuseholders are classified at 600 volts as follows:		
	601-800	
	801-1,200	
	1,201-1,600	
	1,601-2,000	
	2,001-3,000	
	3,001-4,000	
	4,001-5,000	
	5,001-6,000	

There are no 250-volt ratings over 600 amperes, but 600-volt fuses may be used for lower voltages.

7-39 When should AC systems be grounded?

AC circuits and systems shall be grounded as provided for in (a), (b), or (c) in Section 250-5 of the NEC. Other circuits and systems may be grounded:

(a) **Alternating-Current Circuits of less than 50 volts.** AC circuits of less than 50 volts shall be grounded under any of the following conditions:

(1) Where supplied by transformers if the transformer supply system exceeds 150 volts to ground.

(2) Where supplied by transformers if the transformer supply system is ungrounded.

(3) Where installed as overhead conductors outside the building.

(b) **Alternating-current systems of 50 volts and over.** AC systems supplying interior wiring and interior wiring systems shall be grounded under any of the following conditions:

WIRING DESIGN AND PROTECTION

(1) Where the system can be so grounded that the maximum voltage to ground on the ungrounded conductors does not exceed 150 volts.
(2) Where the system is nominally rated 480Y/277-volt, 3-phase, 4-wire in which the neutral is used as a circuit conductor.
(3) Where the system is nominally rated 240/120-volt, 3-phase, 4-wire in which the midpoint of one phase is used as a circuit conductor.
(4) Where a service conductor is uninsulated in accordance with Section 230-4.

(c) **Alternating-Current Systems of 1 KV and over.**
(d) **Separately Derived Systems.**

7-40 When do circuits of 50 volts or less to ground have to be grounded? When do they not have to be grounded?

Circuits of less than 50 volts need not be grounded, unless they are supplied from systems exceeding 150 volts to ground, where they are supplied by transformers from ungrounded systems, or where they run overhead outside buildings (see NEC, Article 250-5a).

7-41 If a grounded conductor is not going to be used in a building, must this grounded conductor be run to each service?

Yes, the grounded conductor must be run to each service. It need not be taken farther than the service equipment, if it will not be required in any of the circuits. This requirement went into effect January 1, 1964 (see NEC, Article 250-23).

7-42 Is it necessary to ground each service on a grounded AC system?

Each individual service must have a ground, and this ground must be connected on the supply side of the service-disconnecting means. If there is only one service from a transformer, there must be an additional ground connection at the transformer or elsewhere in the transformer circuit for the installation to be approved (see NEC, Article 250-23).

7-43 What are the grounding requirements when two or more buildings are supplied from one service?

149

Wiring Design and Protection

Where two or more buildings are supplied from a single service equipment, a grounding electrode at each building shall be connection to the AC system grounded conductor on the supply side of the building disconnecting means of a grounded system or connected to the metal enclosure of the building disconnecting means of an ungrounded system. See Section 250-24 of the NEC.

7-43a Is it required to ground a separately derived system, and if so, where?

Interior wiring systems which are required by Section 250-5 to be grounded, shall be grounded if the phase conductors are not physically connected to another supply system. They shall be grounded at the transformer, generator, or other source of supply, or at the switchboard on the supply side of the disconnecting means. (See NEC, Article 250-26.)

7-45 When must metal, noncurrent-carrying parts of fixed equipment, which are liable to become energized, be grounded?

When supplied by metal-clad wiring, when located in damp or wet places, when within reach of a person who can make contact with a grounded surface or object or within reach of a person standing on the ground, when in hazardous locations, or where any switches, enclosures, etc. are accessible to unqualified persons (see NEC, Article 250-42).

7-46 Is it required to ground metal buildings?

If excessive metal in or on buildings may become energized and is subject to personal contact, adequate bonding and grounding will provide additional safety (see NEC, Article 250-44).

7-47 What is the exception to grounding the noncurrent-carrying parts of portable tools and appliances?

Portable tools and appliances may have double insulation. If they have, they must be so marked and will not require grounding (see NEC, Article 250-45).

7-47a. What are the grounding requirements when lightning rods and conductors are present?

Metal enclosures with conductors must be kept at least 6 feet away from lightning conductors. Should this not be practical, to secure a 6-foot separation, they must be bonded together. (See NEC, Article 250-46).

7-48 What is meant by effective grounding?

The path to ground must be permanent and continuous, must be capable of safely handling the currents that may be imposed on the ground, and must have sufficiently low impedance to limit the potential above ground and to facilitate the opening of the overcurrent devices (see NEC, Article 250-51).

7-49 What is a grounding electrode conductor used for?

The grounding conductor for circuits is used for grounding equipment, conduit, and other metal raceways, including service conduit, cable sheath, etc. (see NEC, Article 250-53).

7-50 What appliances may be grounded to the neutral conductor?

Electric ranges and electric clothes dryers, provided that the neutral is not smaller than No. 10 copper wire (see NEC, Article 250-60).

7-50a Three-wire SE cable with a bare neutral is sometimes used for connecting ranges and dryers. Is it permissible to use this type of cable when the branch circuit originates from a feeder panel?

No, the neutral shall be insulated. (See NEC, Article 250-60.)

7-51 What equipment, other than electric ranges and clothes dryers, may be grounded to the grounding conductor?

The grounding conductor on the supply side of the service-disconnecting means may ground the equipment, meter housing, etc. The load side of the disconnecting means cannot be used for grounding any equipment other than electric ranges and dryers.

7-52 How should continuity at service equipment be assured?

Threaded couplings and bosses should be made wrench tight, as should threadless couplings and connections in rigid conduit or EMT (electrical metallic tubing). Bonding jumpers should be used around concentric and eccentric knockouts (see NEC, Article 250-72).

7-53 On flush-mounted grounded-type receptacles, is it necessary to bond the green grounding screw of the receptacle to the equipment ground?

Yes. (See NEC, Section 250-82(b).)

WIRING DESIGN AND PROTECTION

7-54 Can conduit serve as the equipment ground?

Except for a few cases, such as special precautions in hazardous locations, conduit, armored cable, metal raceways, etc. can serve as the equipment ground.

7-54a What is required to assure electrical continuity of metal raceways and metal-sheathed cable used on voltages exceeding 250 volts?

The electrical continuity of metal raceway or metal-sheathed cable which contains any conductor other than service-entrance conductors or more than 250 volts to ground shall be assured by one of the methods specified in Section 250-72b, (c), (d), and (e), or by one of the following methods: (Refer to Section 250-76 of the NEC.)

(a) Threadless fittings, made up tight, with conduit or metal-clad cable.

(b) Two locknuts, one inside and one outside the boxes and cabinets. Your attention should be specifically called to 250-72d, which calls for bonding to be used around eccentric or concentric knockouts.

7-55 What is the preferred type of grounding electrode?

A metallic underground water system where there are 10 feet or more of buried metallic pipe, including well casings that are bonded to the system, is the preferred grounding electrode (see NEC, Article 250-81).

7-55a Is reinforcing bar in concrete permitted for a grounding electrode, when buried metallic water piping is not available?

Yes. (See NEC Section 250-82d.)

7-56 What is a "made" electrode?

A "made" grounding electrode may be a driven pipe, driven rod, buried plate, or other device approved for the purpose of grounding the equipment and must conform to certain Code requirements (see NEC, Article 250-83).

7-56a What is the No. 1 "made" electrode?

Not less than 20 feet of bare copper conductor of a size specified in Table 250-94a, and in no case smaller than No. 4 AWG,

WIRING DESIGN AND PROTECTION

encased along the bottom of a concrete foundation footing which is in direct contact with the earth. (See NEC, Article 250-83a.)

7-57 What are the requirements for plate electrodes?

They must not have less than 2 square feet of surface exposed to the soil. Electrodes of iron or steel must be at least ¼ inch thick; if of some nonferrous metal, it must be at least 0.06 inch thick (see NEC, Article 250-83).

7-58 What are the requirements for pipe electrodes?

They must be at least ¾ inch trade size and, if made of iron or steel, they must be galvanized or otherwise metal-coated to prevent corrosion; they must also be driven to a depth of at least 8 feet (see NEC, Article 250-83).

7-59 What are the requirements for rod electrodes?

Electrodes of iron or steel must be at least ⅝ inch in diameter; if of nonferrous material, they must be a minimum of ½ inch in diameter. Both types must be driven to a minimum depth of 8 feet (see NEC, Article 250-83).

7-60 Describe the installation of "made" electrodes.

Unless rock bottom is encountered, they should be driven to a minimum depth of 8 feet, and below the permanent-moisture level. Where rock bottom is encountered at a depth of less than 4 feet, they must be buried horizontally in trenches (see NEC, Article 250-83).

7-61 What should be the resistance of "made" electrodes?

Where practical, the resistance of "made" electrodes should not exceed 25 ohms; it is preferred that their resistance be less than this (see NEC, Article 250-84).

7-62 If the electrode resistance is greater than 25 ohms, what should be done?

Additional electrodes should be connected together in parallel and driven to greater depths (see NEC, Article 250-84).

7-63 May grounding conductors be spliced?

No, splices are not permitted; grounding conductors must be one piece for their entire length (see NEC, Article 250-91).

7-64 What is the smallest size grounding conductor permissible?

No. 4 wire or larger may be attached to buildings, No. 6 wire may be used if properly stapled to prevent physical damage,

Wiring Design and Protection

and No. 8 wire may be used if in conduit or armored cable for protection (see NEC, Article 250-92).

7-65 When a common grounding conductor is enclosed in a metallic enclosure, how is the metallic enclosure to be installed?

It shall be electrically continuous from point of attachment to cabinet or enclosure to the point of attachment to the ground clamp or fitting. Metallic enclosures which are not physically and electrically continuous shall be bonded to the grounding conductor at both ends of the metallic enclosure. (See NEC, Article 250-92a.)

7-66 May aluminum be used for a grounding conductor?

Yes, but it may not come in contact with masonry or the earth and cannot be run closer than 18 inches from the earth; also the grounding conductor of a wiring system shall be copper or other corrosion resistant material. (See NEC, Article 250-91a and 250-92.)

7-67 If a water pipe is used as the grounding electrode, what precautions must be taken?

The grounding connection should be made at the point of water-service entrance. If this cannot be done, and there is a water meter on the premises, the water meter should be bonded with a jumper of sufficient length, so that the water meter may be readily removed without disturbing the bonding. The cold water piping should be used, and it should be checked to make certain that there are no insulated connections in the piping (see NEC, Article 250-112).

7-68 Where plastic water pipe serves the house, and a "made" electrode must be used, what grounding procedures are recommended?

It is required that the ground also be bonded to the interior metallic water system, including the hot water piping, and also to the sewer, gas piping, air ducts, etc. This will provide additional safety. (See NEC Article 250-81.)

7-69 When reinforcing bar is used as the grounding electrode, are clamps approved for connecting the grounding conductor to the rebar?

Yes. (See NEC, Section 250-82(b).)

WIRING DESIGN AND PROTECTION

7-70 Can solder be used to attach connections to the grounding conductor?

Solder is never permitted; pressure connectors are to be used (see NEC, Article 250-113).

7-70a Flexible metal conduit is generally not permitted for grounding purposes. Is there any exception to this?

Yes, if not over 6 feet long and protected by overcurrent devices of not over 20 amperes. (See NEC Section 250-91b.)

7-71 What means must be taken to maintain continuity at metallic boxes when nonmetallic systems of wiring are used, so that the equipment ground wire will be continuous?

The equipment grounds must be attached by means of a grounding screw (boxes are now available with a tapped hole having 10-32 threads) or by some other approved means. The equipment ground can be attached by no other means (see NEC, Article 250-114).

7-72 Are sheet-metal straps considered adequate for grounding?

They are not considered adequate, unless attached to a rigid metal base that is seated on the water pipe or other ground electrode (see NEC, Section 250-115).

7-73 What special precautions must be taken in the use of ground clamps?

They must be of a material suitable for use in connection with the material that they are attaching to, because electrolysis may occure if they are made of different metals. All surfaces must be clean and free of paint or corrosion (see NEC, Section 250-115).

7-73a May aluminum grounding conductors be run to cold water piping?

Section 250-115 tells us ground clamps used with aluminum shall be approved for the purpose. At the time of this writing, the author does not know of any approved ground clamps for this purpose. After some research, it was found that cold-water pipes are subject to sweating and moisture and aluminum just do not go together.

WIRING DESIGN AND PROTECTION

7-74 May conductors of different systems occupy the same enclosure?

Conductors for light and power systems of 600 volts or less may occupy the same enclosure; the individual circuits may be AC or DC, but the conductors must all be insulated for the maximum voltage of any conductor in the enclosure (see NEC, Article 300-3a).

7-75 May conductors of systems over 600 volts occupy the same enclosure as conductors carrying less than 600 volts?

No (see NEC, Article 300-3b).

7-76 If the secondary voltage on electric-discharge lamps is 1,000 volts or less, may its wiring occupy the same fixture enclosure as the branch-circuit conductors?

Yes, if insulated for the secondary voltage involved (see NEC, Article 300-3c).

7-77 Is it permissible to run control, relay, or ammeter conductors, which are used in connection with a motor or starter, in the same enclosure as the motor-circuit conductors?

Yes, if the insulation of all the conductors is enough for the highest voltage encountered (see NEC, Article 300-3e).

7-78 May conductors of signal or radio systems be run in the same enclosure as the conductors for light and power?

No, with a few exceptions (see NEC, Article 300-3f).

7-79 Can painted conduit be used outdoors?

No, it is only used indoors (see NEC, Section 300-6).

7-80 When boxes, fittings, conduit, etc. are used in damp or corrosive places, how must they be protected?

They must be protected by a coating of approved corrosion-resistant material (see NEC, Section 300-6).

7-81 In damp locations, what precaution must be taken against the corrosion of boxes, fittings, conduit, etc.?

There must be an air space of at least ¼ inch between the wall or supporting material and fittings, conduit, etc. (see NEC, Section 300-6).

**7-82 When raceways extend from an area of one temper-

Wiring Design and Protection

ature into an area of a widely different temperature, what precautions must be taken?

Precautions must be taken to prevent the circulation of air from a warmer to a colder section through the raceway (see NEC, Section 300-7).

7-83 When raceways and conductors are run through studs, joints, and/or rafters, what precautions must be taken?

They should be run near the approximate center of the wood members, or at least 2 inches from the nearest edge. If the members have to be notched, or a 2-inch protection cannot be given, the conduit or conductors must be covered by a steel plate not less than 1/16 inch in thickness (see NEC, Section 300-4).

7-83a One 240/120 volt multiwire branch circuits feeding through an outlet box, may the neutral be broken and connected by means of the screws on the receptacle?

No. In multiwire circuits, the continuity of an unidentified grounded conductor shall not be dependent upon, as a connection device, such as lampholders, receptacles, etc., where the removal of such devices would interrupt the continuity. (See NEC Section 300-13.)

7-84 How much wire must be allowed at outlets and switch boxes for connections and splices?

There must be at least 6 inches of free conductor left for making the connections (see NEC, Article 300-14).

7-85 What precautions must be taken to prevent induced currents in metal enclosures?

When conductors carry AC, all phase wires and the neutral wire, if one is used, must run in the same raceway. When single conductors must be passed through metal having magnetic properties (iron and steel), slotting the metal between the holes will help keep down the inductive effect (See NEC, Article 300-20).

7-86 When it is necessary to run wires through air handling ducts or plenums, what precautions must be taken?

The conductors must be in conduit, electrical metallic tubing, flexible steel conduit with lead-covered conductors, Type ACL metal-clad cable, Type MI cable, or Type ALS cable (see NEC, Article 300-22).

WIRING DESIGN AND PROTECTION

7-86a Does temporary wiring, such as for construction, Christmas lighting, carnivals, etc., come under the NEC?
Yes, a new Section 305-1(b) appears in the 1975 NEC.

7-86b Are ground-fault circuit-interrupters require on temporary wiring for construction sites?
Yes; it will be effective January 1, 1974 on all 15-and 20-ampere branch circuits. (See NEC Section 210-7.)

7-87 When conductors are used underground in concrete slabs or other masonry that comes in direct contact with the earth, or where condensation or accumulated moisture in raceways is apt to occur, what characteristic must the insulation have?
It must be moisture-resistant (see NEC, Article 310-5).

7-88 Name some moisture-resistant insulations.
RHW, RUW, TW, THW, THWN, and XHHW type cable.

7-89 What are the cable requirements for buried conductors?
Cables of one or more conductors for direct burial in the earth must be Type USE cable, except for branch circuits and feeders, which may use Type UF cable. Type UF cable cannot be used for service wires (see NEC, Section 310-5).

7-90 What other rules apply to buried conductors?
All conductors, including the neutral, must be buried in the same trench and be continuous (without any splice). Extra mechanical protection may be required, such as a covering board, concrete pad, or raceway. These are not required by the NEC but may be legally required by the inspection authority (see NEC, Section 300-5 and 310-5).

7-91 What is the minimum size of conductors allowed by the NEC?
No. 14 wire; however, there are exceptions where smaller sizes may be used (see NEC, Section 310-5).

7-92 What is the ruling on stranded conductors?
Except for bus bars and Type MI cable, conductors of No. 6 and larger must be stranded (see NEC, Section 310-3).

This requirement will be changed to No. 8 and larger effective January 1, 1973.

WIRING DESIGN AND PROTECTION

7-93 What is the minimum size of conductors that may be paralleled?

Conductors in sizes 1/0 and larger may be run in multiple (see NEC, Section 310-4).

7-94 When conductors are run in multiple, what factors must be considered?

They must be the same length and of the same conductor material; have the same circular-mil area and the same type of insulation; and be arranged to terminate at both ends so that there will be equal distribution of current between the conductors. Phases must be coordinated to eliminate any induction currents that may be caused to flow in the raceways (see NEC, Section 310-4).

7-95 How many conductors may be run in raceways or cables without having to apply a derating factor to the current-carrying capacity?

Three conductors; when more than three current-carrying conductors are run, a derating factor must be applied (see NEC, Notes to Table 310-16 through 310-19).

7-96 What are the derating factors for more than three conductors in raceways or cables?

4 to 6 conductors .... 80% rating
7 to 24 conductors .... 70% rating
25 to 42 conductors .... 60% rating
43 conductors and above .... 50% rating
(see NEC, Notes to Table 310-16 through 310-19).

7-97 In derating, how is the neutral conductor considered?

Normally the current in the neutral conductor is only the unbalanced current; therefore, if the system is well balanced, the neutral is not considered as a current-carrying conductor and would not enter into derating (See NEC, Note 10 to Table 310-16 through 310-19).

7-98 In determining the current in the neutral conductor of a Wye system, how is the neutral classed?

In a 4-wire, 3-phase, Wye-connected system, a common conductor carries approximately the same current as the other conductors and is therefore not considered as a neutral in determin-

WIRING DESIGN AND PROTECTION

Table 2. Correction Factors Ambient Temps. Over 30°C. 86°F.

C.	F.	60°C (140°F)	75°C (167°F)	85°C (185°F)	90°C (194°F)	110°C (230°F)	125°C (257°F)	200°C (392°F)	250°C (482°F)
40	104	.82	.88	.90	.90	.94	.95	....	....
45	113	.71	.82	.85	.85	.90	.92	....	....
50	122	.58	.75	.80	.80	.87	.89	....	....
55	131	.41	.67	.74	.74	.83	.86	....	....
60	140	....	.58	.67	.67	.79	.83	.91	.95
70	158	....	.35	.52	.52	.71	.76	.87	.91
75	167	....	....	.43	.43	.66	.72	.86	.89
80	176	....	....	.30	.30	.61	.69	.84	.87
90	194	....	....	....	....	.50	.61	.80	.83
100	212	....	....	....	....	....	.51	.77	.80
120	248	....	....	....	....	....	....	.69	.72
140	284	....	....	....	....	....	....	.59	.59
160	320	....	....	....	....	....	....	....	.54
180	356	....	....	....	....	....	....	....	.50
200	392	....	....	....	....	....	....	....	.43
225	437	....	....	....	....	....	....	....	.30

Table 3. Allowable Ampacities of Insulated Copper Conductors

Not More than Three Conductors in Raceway or Cable or Direct Burial (Based on Ambient Temperature of 30°C. 86°F.)

Size AWG MCM	60°C (140°F) TYPES RUW (14-2), T, TW	75°C (167°F) TYPES RH, RHW, RUH (14-2), THW, THWN, XHHW	85°C (185°F) TYPES V, MI	90°C (194°F) TYPES TA, TBS, SA, AVB SIS, FEP, FEPB, RHH, THHN, XHHW**	110° (230°F) TYPES AVA, AVL	125°C (257°F) TYPES AI (14-8), AIA	200°C (392°F) TYPES A (14-8), AA, FEP* FEPB*	250°C (482°F) TYPE TFE (Nickel or nickel-coated copper only)
14	15	15	25	25†	30	30	30	40
12	20	20	30	30†	35	40	40	55
10	30	30	40	40†	45	50	55	75
8	40	45	50	50	60	65	70	95

ing the derating of current capacity (see NEC, Note 10 to Table 310-16 through 310-19).

7-99 Give some current-carrying capacities of insulated copper conductors. (See Table 3.)

160

Wiring Design and Protection
Table 3. Allowable Ampacities of Insulated Copper Conductors
Not More than Three Conductors in Raceway or Cable or Direct Burial (Based on Ambient Temperature of 30°C. 86°F.)

Size AWG MCM	60°C (140°F) TYPES RUW (14-2), T, TW	75°C (167°F) TYPES RH, RHW, RUH (14-2), THW, THWN, XHHW	85°C (185°F) TYPES V, MI	90°C (194°F) TYPES TA, TBS, SA, AVB SIS, FEP, FEPB, RHH, THHN, XHHW**	110° (230°F) TYPES AVA, AVL	125°C (257°F) TYPES AI (14-8), AIA	200°C (392°F) TYPES A (14-8), AA, FEP*, FEPB*	250°C (482°F) TYPE TFE (Nickel or nickel-coated copper only)
6	55	65	70	70	80	85	95	120
***4	70	85	90	90	105	115	120	145
***3	80	100	105	105	120	130	145	170
***2	95	115	120	120	135	145	165	195
***1	110	130	140	140	160	170	190	220
***0	125	150	155	155	190	200	225	250
***00	145	175	185	185	215	230	250	280
000	165	200	210	210	245	265	285	315
0000	195	230	235	235	275	310	340	370
250	215	255	270	270	315	335	....	....
300	240	285	300	300	345	380	....	....
350	260	310	325	325	390	420	....	....
400	280	335	360	360	420	450	....	....
500	320	380	405	405	470	500	....	....
600	355	420	455	455	525	545	....	....
700	385	460	490	490	560	600	....	....
750	400	475	500	500	580	620	....	....
800	410	490	515	515	600	640	....	....
900	435	520	555	555	....	....	....	....
1000	455	545	585	585	680	730	....	....
1250	495	590	645	645	....	....	....	....
1500	520	625	700	700	785	....	....	....
1750	545	650	735	735	....	....	....	....
2000	560	665	775	775	840	....	....	....

* Special use only. See Table 310-2(a).
** For dry locations only. See Table 310-2(a).
These ampacities relate only to conductors described in Table 310-2(a).
*** For 3-wire, single-phase residential services, the allowable ampacity of RH, RHH, RHW, THW and XHHW copper conductors shall be for sizes No. 4-100 Amp., No. 3-110 Amp., No. 2-125 Amp., No. 1-150 Amp., No. 1/0-175 Amp., and No. 2/0-200 Amp.
† The ampacities for Types FEP, FEPB, RHH, THHN, and XHHW conductors for sizes AWG 14, 12 and 10 shall be the same as designated for 75°C conductors in this Table. For ambient temperatures over 30°C, see Correction Factors, Note 13.

WIRING DESIGN AND PROTECTION

Table 4. Allowable Ampacities of Insulated Aluminum and Copper-Clad Aluminum Conductors

Not More than Three Conductors in Raceway or Cable or Direct Burial (Based on Ambient Temperature of 30°C. 86°F.)

Size AWG MCM	60°C (140°F) TYPES RUW (12-2), T, TW	75°C (167°F) TYPES RH, RHW, RUH (12-2), THW, THWN XHHW	85°C (185°F) TYPES V, MI	90°C (194°F) TYPES TA, TBS, SA, AVB, SIS, RHH THHN XHHW**	110°C (230°F) TYPES AVA, AVL	125°C (257°F) TYPES AI (12-8), AIA	200°C (392°F) TYPES A (12-8), AA
12	15	15	25	25†	25	30	30
10	25	25	30	30†	35	40	45
8	30	40	40	40	45	50	55
6	40	50	55	55	60	65	75
4	55	65	70	70	80	90	95
3	65	75	80	80	95	100	115
*2	75	90	95	95	105	115	130
*1	85	100	110	110	125	135	150
*0	100	120	125	125	150	160	180
*00	115	135	145	145	170	180	200
*000	130	155	165	165	195	210	225
*0000	155	180	185	185	215	245	270
250	170	205	215	215	250	270	....
300	190	230	240	240	275	305	....
350	210	250	260	260	310	335	....
400	225	270	290	290	335	360	....
500	260	310	330	330	380	405	....
600	285	340	370	370	425	440	....
700	310	375	395	395	455	485	....
750	320	385	405	405	470	500	....
800	330	395	415	415	485	520	....
900	355	425	455	455	....	....	....
1000	375	445	480	480	560	600	....
1250	405	485	530	530	....	....	....
1500	435	520	580	580	650	....	....
1750	455	545	615	615	....	....	....
2000	470	560	650	650	705	....	....

These ampacities relate only to conductors described in Table 310-2(a).
  * For 3-wire, single-phase residential services, the allowable ampacity of RH, RHH, RHW, THW, and XHHW conductors shall be for sizes No. 2-100 Amp., No. 1-110 Amp., No. 1/0-125 Amp., No. 2/0-150 Amp., No. 3/0-175 Amp. and No. 4/0-200 Amp.

162

WIRING DESIGN AND PROTECTION

** For dry locations only. See Table 310-2(a).
† The ampacities for Type RHH, THHN, and XHHW conductors for sizes AWG 12 and 10 shall be the same as designated for 75°C conductors in this Table.
For ambient temperatures over 30°C, see Correction Factors, Note 13.

7-100 Give some current-carrying capacities of insulated aluminum conductors. (See Table 4.)

7-101 What advantages would two parallel 500-MCM cables have over one 1,000-MCM cable?

They would be easier to handle and to pull into raceways. According to Table 2, 1,000-MCM Type RH wire has a current-carrying capacity of 545 amperes, whereas 500-MCM cable has a current-carrying capacity of 380 amperes; therefore, two 500-MCM cables in parallel would have the same circular-mil area as one 1,000-MCM cable and a current-carrying capacity of 760 amperes as against the 545 amperes current-carrying capacity of one 1,000-MCM cable. This is approximately 40% more current-carrying capacity for the two 500-MCM.

7-102 May THW insulated conductors be run through a continuous row of fluorescent fixtures?

Yes; see NEC Table 310-13.

7-103 In figuring the ampacity of copper-clad aluminum conductors, is Table 2 or 4 used?

Table 4 is used. The ampacity is the same as the same size conductor of aluminum only. (See NEC, Table 310-14 or Table 4 in this book.)

CHAPTER 8

Wiring Methods and Materials

8-1 Can cable trays supports be made of combustible material?
No, they must be made of noncombustible material and may be ventilated or unventilated (see NEC, Article 318-1).

8-2 Are cable trays supports intended to be used with ordinary conductors (as described in Article 310 of the NEC)?
No, they are not intended for this purpose (see NEC, Article 318-1).

8-3 Do cable trays have to be used as a complete system, or may they be used as only partial cable supports?
They must be used as a complete system, including boxes and fittings, if used; if necessary, noncombustible covers or enclosures may be required (see NEC, Section 318-5).

8-4 Is grounding required on metallic cable trays?
They must be grounded but must not be used as either the neutral conductor or as an equipment ground (see NEC, Section 318-6).

8-6 Is open wiring on insulators, commonly known as knob-and-tube wiring, approved by the National Electrical Code?
Yes, although you may find that local inspection authorities and regulations may not approve it (see NEC, Section 320-3 and 324-3).

165

Wiring Methods and Materials

8-7 What is Type MI cable?

Type MI cable is a cable in which one or more electrical conductors are insulated with a highly compressed refractory mineral insulation (magnesium oxide) and enclosed in a liquid-tight and gas-tight metallic sheathing (copper) (see NEC, Article 330-1).

8-8 Where can Type MI cable be used?

This is one wiring material that can be used for practically every conceivable type of service or circuit; when it is exposed to cinder fill or other destructive corrosive conditions, it must be protected by materials suitable for these conditions (see NEC, Section 330-3).

8-9 What precautions must be taken with Type MI cable to prevent the entrance of moisture?

Where Type MI cable terminates, an approved seal must be provided immediately after stripping to prevent the entrance of moisture into the mineral insulation; the conductors must be insulated with an approved insulation where they extend beyond the sheath (see NEC, Section 330-15).

8-11 Does the outer sheath of Type MI cable meet the requirements for equipment grounding?

Yes; it provides an adequate path for equipment grounding purposes but not for use as a grounded or neutral conductor (see NEC, Section 250-57(a) and 250-91(b)).

8-12 What types of cable are considered metal-clad cable?

Both MC- and AC-type cables (NEC 334-4).

8-13 What is Type MC cable?

Type MC cables are power and control cables in the size range from No. 14 and larger for copper and No. 12 and larger for aluminum and copper-clad aluminum. For conductor sizes and insulation thickness of Type MC Cable rated at over 600 volts, see Article 310, Part C, of the NEC.

The metal enclosures are either a covering of interlocking metal tape or an approved, impervious, close-fitting, corrugated tube. Supplemental protection of an outer covering of corrosion-resistant material is required where such protection is needed (see NEC, Article 334-4).

8-14 Does Type MC cable provide an adequate path for grounding purposes?

WIRING METHODS AND MATERIALS

Yes (see NEC, Article 334-4).

8-15 What is Type AC cable?

Type AC cables are branch-circuit and feeder cables with an armor of flexible metal tape. Cable of the AC type, except Type ACL, must have an internal bonding strip of copper or aluminum, in tight physical contact with the armor for its entire length (see NEC, Article 334-4).

8-16 At what intervals must Type MC cable be supported?

At intervals not exceeding 6 feet and within 2 feet of every box or fitting, except where the cable is fished (see NEC, Article 334-8).

8-17 At what intervals must Type AC cable be supported?

At intervals of not more than 4½ feet and within 12 inches of boxes and fittings, except where the cable is fished and except lengths of not over 24 inches at terminals where flexibility is necessary (see NEC, Article 334-8).

8-18 How may bends be made in Types MC and AC cables?

Bends must be made so that the cables are not injured. The radius of the curve of the inner edge of any bend must not be less than 7 times the diameter of Type MC cable nor 5 times the diameter of Type AC cable (see NEC, Article 334-9).

8-19 What extra precaution is necessary with Type AC cable when attaching fittings?

An approved insulating bushing or its equivalent approved protection must be provided between the conductors and the armor (see NEC, Article 334-10).

8-20 What is nonmetallic sheathed cable?

Nonmetallic sheathed cable is an assembly of two or more insulated conductors having an outer sheath of moisture-resistant, flame-retardant, nonmetallic material (see NEC, Article 336-1).

8-21 What are the available sizes of nonmetallic sheathed cable?

Sizes No. 14 to No. 2 (AWG) inclusive (see NEC, Article 336-2).

8-22 What is the difference between Type NM and Type NMC cables?

In addition to having flame-retardant and moisure-resistant covering, Type NMC cable must also be fungus-resistant and corrosion-resistant (see NEC, Article 336-2).

Wiring Methods and Materials

8-23 Where may Type NM cable be used?

It may be used for exposed and concealed work in normally dry locations. It may also be installed or fished in the air voids of masonry block or tile walls, where such walls are not exposed to excessive moisture or dampness (see NEC, Article 336-3).

8-23a On dwelling occupancies, what if any restrictions have been added?

Types NM and NMC cables shall be permitted to be used in one and two family dwellings or multifamily dwellings and other structures not exceeding three floors above grade. (See NEC, Section 336-3).

8-24 Where may Type NMC cable be installed?

Where exposed to corrosive fumes or vapors, for exposed and concealed work, in dry and moist or damp places, inside or outside of masonry block or tile walls. It may also be imbedded in a 2-inch chase and covered with plaster in masonry walls, but it must be protected from nails by a covering of corrosion-resistant steel at least 1/16 inch thick (see NEC, Article 336-3).

8-25 Where is it not permissible to use Type NM or Type NMC cable?

As service-entrance cable, in commercial garages, in theaters, and similar locations, except as provided in Article 518, "Places of Assembly."

8-26 What supports are necessary for Types NM and NMC cable?

The installation must not be subject to injury, and the cable must be secured in place at intervals not exceeding 4½ feet and supported within 12 inches from boxes and fittings (see NEC, Article 336-5).

8-26a. What is Shielded Nonmetallic Sheathed Cable?

Type SNM, shielded nonmetallic-sheathed cable, is a factory assembly of two or more insulated conductors in an extruded core of moisture-resistant, flame-retardant nonmetallic material, covered with an overlapping spiral metal tape and wire shield and jacketed with an extruded moisture, flame, oil, corrosion, fungus and sunlight-resistant nonmetallic material. (See NEC, Section 337-1.)

8-26b Where is SNM cable permitted to be used?

Type SNM cable may be used only as follows:
1. Where operating temperatures do not exceed the rating marked on the cable.
2. In continuous rigid cable supports in raceways.
3. In hazardous locations where permitted in Articles 500 through 516. (See NEC, Section 337-3.)

8-27 What is service-entrance cable?

Service-entrance cable is a conductor assembly provided with a suitable covering, primarily used for services (see NEC, Article 338-1).

8-28 What are the two types of service-entrance cable?

Type SE, and Type USE. (See NEC, Section 338-1.)

8-30 What is Type SE cable?

It has a flame-retardant, moisture-resistant covering but is not required to have an inherent protection agaiust mechanical abuse (See NEC, Article 338-1).

8-31 What is Type USE cable?

For underground use, it has a moisture-resistant covering but is not required to have a flame-retardant covering or an inherent protection against mechanical abuse (see NEC, Article 338-1).

8-32 Can service-entrance cable be used as branch-circuit conductors or feeder conductors?

It may be used for interior wiring systems only when all circuit conductors of the cable are of the rubber-covered or thermoplastic type. However, service-entrance cable without individual insulation on the grounded conductor can be used to supply ranges or clothes dryers, when the cable does not have an outer metallic covering and does not exceed 150 volts to ground. It may also be used as a feeder when it supplies other buildings on the same premises. SE cable with a bare neutral shall not originate in a feeder panel, as covered in section 250-60. (See NEC, Article 338-3.)

8-33 Describe underground feeder and branch-circuit cable.

Underground feeder and branch-circuit cable must be an approved Type UF cable in sizes No. 14 to No. 4/0 AWG, inclusive. The ampacity of Type UF cable shall be that of 60°C (140°F) conductors in Tables 310-16 and 310-18. In addition, the cable may include a bare conductor for grounding purpose only. The overall covering must also be approved for direct burial in the earth (see NEC, Article 339-1).

8-34 What is required on overload protection for Type UF cable?

Underground feeder and branch-circuit cable may be used underground, including direct burial, provided they have overcurrent protection that does not exceed the capacity of the conductors (see NEC, Article 339-3).

8-35 May conductors of the same circuit be run in separate trenches?

All conductors of the same circuit must be run in the same trench or raceway (see NEC, Article 339-3).

8-36 What depth must UF cable be buried?

A minimum of 24 inches (see NEC, Table 300-5).

8-37 May UF cable be used for interior wiring?

Yes; however, it must not be exposed to direct sunlight unless it is specifically approved for this purpose. Many UF cables are subject to damage or deterioration under exposure to sunlight (see NEC, Article 339-3).

8-38 What is the normal taper on a standard conduit thread-cutting die?

¾ inch per foot (see NEC, Article 346-1).

8-39 Where can rigid metal conduit be used?

Practically everywhere; however, rigid metal conduit protected solely by enamel cannot be used outdoors. Also, where practicable, dissimilar metals should not come in contact anywhere in the system to avoid the possibility of galvanic action, which could prove destructive to the system. (See NEC, Section 346-1.)

8-40 When subject to cinder fill, what precautions must be taken with rigid metal conduit?

It must be protected by a minimum of 2 inches of noncinder concrete (see NEC, Article 346-3).

8-41 What is the minimum size of rigid metal conduit?

½-inch trade size (see NEC, Article 346-5). (With a few exceptions.)

8-42 What are the requirements for couplings on rigid metal conduit?

WIRING METHODS AND MATERIALS

Threadless couplings and connectors must be made tight. Where installed in wet places or buried in concrete masonry or fill, the type used must prevent water from entering the conduit (see NEC, Article 346-9).

8-43 Are running threads permitted on rigid metal conduit?

They cannot be used on conduit for connections at couplings (see NEC, Article 346-9).

8-44 What is the maximum number of bends permitted in rigid metal conduit between outlets?

Not more than 4 quarter bends or a total of 360°; this includes offsets, etc. at outlets and fittings (see NEC, Article 346-11).

8-45 What is the minimum radius of conduit bends?

See Tables 1 and 1A. Table 1 is for regular field bends. Table 1A is for field bends made with a one-shot bender.

Table 1. Radius of Conduit Bends

Size of Conduit	Conductors Without Lead Sheath	Conductors With Lead Sheath
½ in.	4 in.	6 in.
¾ in.	5 in.	8 in.
1 in.	6 in.	11 in.
1¼ in.	8 in.	14 in.
1½ in.	10 in.	16 in.
2 in.	12 in.	21 in.
2½ in.	15 in.	25 in.
3 in.	18 in.	31 in.
3½ in.	21 in.	36 in.
4 in.	24 in.	40 in.
5 in.	30 in.	50 in.
6 in.	36 in.	61 in.

8-46 After a cut is made in rigid metal conduit, what precaution must be taken?

All cut ends of conduits must be reamed to remove any rough ends which might damage the wire when it is pulled in (see NEC, Article 346-7).

8-46a What number of conductors are permitted in conduit on new work and on rewire work?

WIRING METHODS AND MATERIALS

The number of conductors is now the same for new work and rewire work. The 1971 NEC came out with new Table applicable to both new and rewire fills for conductor. See NEC Tables in Chapter 9.

Table 1A. Radius of Conduit Bends

Size of Conduit (inches)	Radius to Center of Conduit (inches)
½ in.	4 in.
¾ in.	4½ in.
1 in.	5¾ in.
1¼ in.	7¼ in.
1½ in.	8¼ in.
2 in.	9½ in.
2½ in.	10½ in.
3 in.	13 in.
3½ in.	15 in.
4 in.	16 in.
4½ in.	20 in.
5 in.	24 in.
6 in.	30 in.

8-47 Should rigid nonmetallic conduit be of an approved type?

Yes.

8-48 Is plastic water pipe approved to be used as conduit?

It may not be used.

8-49 What uses are permitted for rigid nonmetallic conduit?

For 600 volts or less, except for direct burial where it is not less than 18 inches below grade and where the voltage exceeds 600 volts, it must then be encased in 2 inches of concrete; in concrete walls, floors and ceilings; in locations where subject to severe corrosive influences; in cinder fill; and in wet locations (see NEC, Article 347-2).

8-50 Where is the use of nonmetallic rigid conduit prohibited?

In hazardous locations, in concealed spaces of combustible construction, for the support of fixtures or other equipment, or where subject to physical damage (see NEC, Article 347-3).

8-51 Where must expansion joints be used on rigid non-

WIRING METHODS AND MATERIALS

metallic conduit?

Where required to compensate for thermal expansion and contraction (see NEC, Article 347-9).

8-52 What is the minimum size of nonmetallic rigid conduit?

No conduit smaller than ½-inch electrical trade size may be used (see NEC, Article 347-10).

8-53 Are bushings required when rigid nonmetallic conduit is used?

Yes (see NEC, Article 347-12).

8-54 What is electrical metallic tubing (EMT)?

A thin-walled metal conduit.

8-55 May EMT be threaded?

No; EMT can be coupled only by approved fittings (see NEC, Article 348-7).

8-56 What are the minimum and maximum sizes that EMT is available in?

The minimum size is ½ inch, and the maximum size is 4 inches, electrical trade size (see NEC, Article 348-5).

8-57 May more or less conductors be contained in EMT than in rigid metal conduit?

The same number of wires apply to both conduits (see NEC, Article 346-6).

8-58 How do the number of bends and the reaming requirements for EMT compare to rigid metal conduit?

They are the same for both (see NEC, Articles 348-10 and 348-11).

8-59 When EMT is exposed to moisture or used outdoors, what type of fittings must be used?

They must be rain-tight fittings (see NEC, Article 348-8).

8-60 When EMT is buried in concrete, what type fittings must be used?

They must be concrete-tight fittings (see NEC, Article 348-8).

8-61 What are some common names for flexible metal conduit?

Flex, Greenfield.

8-62 What is the minimum size for flexible metal conduit?

The normal minimum size is ½ inch, although there are some exceptions to this (see NEC, Article 350-3).

173

WIRING METHODS AND MATERIALS

8-63 What is the number of wires permitted in flexible metal conduit?

The requirement for this is the same as for rigid metal conduit; however, when ⅜-inch conduit is permitted, a special conductor table is required (see Table 2).

Table 2. Maximum Number of Insulated Conductors in ⅜ In. Flexible Metal Conduit.*

Col. A = With fitting inside conduit.
Col. B = With fitting outside conduit.

Size AWG	Types RF-2, RFH-2, SF-2		Types TF, T, XHHW, AF, TW, RUH, RUW		Types TFN, THHN, THWN		Types FEP, FEPB, PF, PGF	
	A	B	A	B	A	B	A	B
18	..	3	3	7	4	8	5	8
16	..	2	2	4	3	7	4	8
14	..	..	..	4	3	7	3	7
12	..	..	..	3	..	4	..	4
10	..	..	..	..	..	2	..	3

* In addition one uninsulated grounding conductor of the same AWG size may be installed.

8-63a What are the approved sizes of liquidtight flexible metal conduit?

The sizes of liquidtight flexible metal conduit shall be electrical trade sizes ½ to 4 inches inclusive.

Exception: ⅜-inch size may be used as permitted in Section 350-3.

8-63b How often should liquidtight flexible metal conduit be supported?

Where liquidtight flexible metal conduit is installed as a fixed raceway, it shall be secured by approved means at intervals not exceeding 4½ feet and within 12 inches on each side of every outlet box or fitting except where conduct is fished. (See NEC, section 351-6.)

8-63c May regular flexible metal conduit fittings be used with liquidtight flexible metal conduit?

No, they shall be fittings approved for use with liquidtight flexible metal conduit. (See NEC Section 351-5.)

8-63d May liquidtight flexible metal conduit be used as an equipment ground?

WIRING METHODS AND MATERIALS

Liquidtight flexible metal conduit may be used for grounding in the 1¼ inch and smaller trade sizes if the length is 6 feet or less and it is terminated in fittings approved for the purpose. (See NEC, Section 351-7.)

8-64 What are wireways?

Wireways are sheet-metal troughs with hinged or removable covers for housing and protecting electrical wires and cable and in which conductors are laid in place after the wireway has been installed as a complete system (see NEC, Article 362-1).

8-65 What use of wireways is permitted?

Wireways may be installed only for exposed work; where wireways are intended for outdoor use, they must be of approved, rain-tight construction (see NEC, Article 362-2).

8-66 What are the uses for which wireways are prohibited?

Where subject to severe physical damage or corrosive vapors and in hazardous locations (see NEC, Article 362-2).

8-67 What is the largest size wire permitted in wireways?

No conductor larger than that for which the wireway is designed shall be installed in any wireway (see NEC, Section 362-4).

8-68 What is the maximum number of conductors permitted in wireways, and what is the maximum percentage of fill?

A maximum of 30 conductors at any cross section; signal circuits or starter-control wires are not included. The maximum fill cannot exceed 20% of the interior cross-sectional area (see NEC, Article 362-5).

8-69 Are splices permitted in wireways?

Yes, provided that the conductors with splices do not take up more than 75% of the area of the wireway at any point (see NEC, Article 362-6).

8-70 May the wireways be open at the ends?

No, the dead ends must be closed (see NEC, Article 362-9).

8-70a Is the grounding conductor used in figuring the number of conductors allowed in a box?

Yes; but only one conductor is to be counted. (See NEC, Section 370-6.)

8-71 What is the purpose of auxiliary gutters?

Wiring Methods and Materials

They are used to supplement wiring spaces at meter centers, distribution centers, switchboards, and similar points of wiring systems; they may enclose conductors or bus bars but cannot be used to enclose switches, overcurrent devices, or other appliances or apparatus (see NEC, Article 374-1).

8-72 What is the maximum number of conductors permitted in auxiliary gutters, and what is the maximum percentage of fill allowed?

A maximum of 30 conductors at any cross section; signal circuits and starter control wires are not included. The maximum fill cannot exceed 20% of the interior cross-sectional area (see NEC, Article 374-5).

8-73 What is the current-carrying capacity of conductors in auxiliary gutters?

The ampacities of insulated copper and aluminum conductors are given in Tables 3 and 4 respectively. When the number of current-carrying conductors contained in the auxiliary gutter are 30 or less, the correction factors specified in Note 8 of these Tables shall not apply. The current-carried continuously in bare copper bars in auxiliary gutters shall not exceed 1000 amperes per square inch of cross section of the conductor. For aluminum bars the current carried continuously shall not exceed 700 amperes per square inch of cross section of the conductor. (See NEC, Section 374-5.)

8-74 May taps and splices be made in auxiliary gutter?

Yes; however, splices may not occupy more than 75% of the cross-sectional area at any point. All taps must be identified as to the circuit or equipment that they supply (see NEC, Article 374-8).

8-75 May round boxes be used as outlet, switch, or junction boxes?

No, they cannot be used where conduits or connectors with locknuts or bushings are used (see NEC, Article 370-2).

8-76 Is the number of conductors permitted in outlet, switch, and junction boxes limited to any certain number?

Yes; see Tables 3 and 5.

Where there is not sufficient space for a deeper box, four No. 14 AWG conductors may enter a box provided with cable clamps and

WIRING METHODS AND MATERIALS

containing one or more devices on a single mounting strap.

Any box less than 1½ inches deep is considered to be a shallow box.

These tables are quite plain; however, they are for conduit or where connectors are used. If cable clamps are used, one conductor

Table 3. Boxes

Box Dimension, Inches Trade Size or Type	Min. Cu. In. Cap.	Maximum Number of Conductors				
		#14	#12	#10	#8	#6
4 × 1¼ Round or Octagonal	12.5	6	5	5	4	0
4 × 1½ Round or Octagonal	15.5	7	6	6	5	0
4 × 2⅛ Round or Octagonal	21.5	10	9	8	7	0
4 × 1¼ Square	18.0	9	8	7	6	0
4 × 1½ Square	21.0	10	9	8	7	0
4 × 2⅛ Square	30.3	15	13	12	10	6*
4¹¹⁄₁₆ × 1¼ Square	25.5	12	11	10	8	0
4¹¹⁄₁₆ × 1½ Square	29.5	14	13	11	9	0
4¹¹⁄₁₆ × 2⅛ Square	42.0	21	18	16	14	6
3 × 2 × 1½ Device	7.5	3	3	3	2	0
3 × 2 × 2 Device	10.0	5	4	4	3	0
3 × 2 × 2¼ Device	10.5	5	4	4	3	0
3 × 2 × 2½ Device	12.5	6	5	5	4	0
3 × 2 × 2¾ Device	14.0	7	6	5	4	0
3 × 2 × 3½ Device	18.0	9	8	7	6	0
4 × 2⅛ × 1½ Device	10.3	5	4	4	3	0
4 × 2⅛ × 1⅞ Device	13.0	6	5	5	4	0
4 × 2⅛ × 2⅛ Device	14.5	7	6	5	4	0
3¾ × 2 × 2½ Masonry Box/gang	14.0	7	6	5	4	0
3¾ × 2 × 3½ Masonry Box/gang	21.0	10	9	8	7	0
FS—Minimum Internal Depth 1¾ Single Cover/Gang	13.5	6	6	5	4	0
FD—Minimum Intenal Depth 2⅜ Single Cover/Gang	18.0	9	8	7	6	3
FS—Minimum Internal Depth 1¾ Multiple Cover/Gang	18.0	9	8	7	6	0
FD—Minimum Internal Depth 2⅜ Multiple Cover/Gang	24.0	12	10	9	8	4

* Not to be used as a pull box. For termination only.

must be deducted from the table to allow for the clamp space. Where one or more fixture studs, cable clamps, or hickeys are contained in the box, the number of conductors will be one less than shown in the tables, with a further deduction of one conductor

Wiring Methods and Materials

for one or several flush devices mounted on the same strap. A double receptacle is one device, a single switch is one device, three despard switches on one strap is one device, three despard receptacles on one strap is one device.

There are times when you will have combinations of wire sizes or other conditions, and you might not find the answer in Tables 3 or 4. The following table gives the volume of free space (in cubic inches) within a box that is required for each wire size specified. Determine the number of cubic inches in the box; from this you can find the number of various combinations of wire (see NEC, Article 370-6).

8-77 What protection must be given to conductors entering boxes or fittings?

They must be protected from abrasion, and the openings through which conductors enter must be adequately closed (see NEC, Article 370-7).

Table 5.

Size of Conductor	Free Space Within Box for Each Conductor
No. 14	2.00 cubic inches
No. 12	2.25 cubic inches
No. 10	2.50 cubic inches
No. 8	3.00 cubic inches
No. 6	5.00 cubic inches

8-78 May unused openings in boxes and fittings be left open?

No, they must be adequately closed with protection equivalent to that of the wall of the box or fitting (see NEC, Article 370-8).

8-79 How far may boxes be set back in walls?

In walls and ceilings of concrete or other noncombustible materials, the boxes may be set back not more than ¼ inch. In walls or ceilings of wood or other combustible materials, the boxes must be flush with the finished surface. If the plaster is broken or incomplete, it must be repaired, so that there will be no gaps or openings around the box (see NEC, Articles 370-10 and 370-11).

8-80 How can you determine dimensions for pull or junction boxes?

WIRING METHODS AND MATERIALS

For raceways of 1 inch trade size and larger containing conductors of No. 6 wire or larger, the minimum dimensions of a pull or junction box installed in the raceway for straight pulls must be not less than 8 times the trade diameter of the largest raceway. For angle or U pulls, the distance between each raceway entry inside the box and opposite wall of the box must not be less than 6 times the trade size of the raceway. This distance must be increased for additional entries by the amount of the sum of the diameters of all other raceway entries on the same wall of the box. The distance between raceway entries enclosing the same conductor cannot be less than 6 times the electrical trade diameter of the larger raceway (see NEC, Article 370-18).

8-81 Must pull, outlet, and junction boxes be accessible?

Yes, they must be accessible without removing any part of the building, sidewalk, or paving (see NEC, Article 370-19).

8-82 When installing cabinets and cutout boxes in damp or wet places, what precautions must be taken?

They must be installed so that any accumulated moisture will drain out. There must also be a ¼-inch air space between the

Table 6. Minimum Wire Bending Space at Terminals and Minimum Width of Wiring Gutters in Inches

AWG or Circular-Mil Size of Wire	1	2	3	4	5
14-8	Not Specified	—	—	—	—
6	1½	—	—	—	—
4 - 3	2	—	—	—	—
2	2½	—	—	—	—
1	3	—	—	—	—
0 - 00	3½	5	7	—	—
000 - 0000	4	6	8	—	—
250 MCM	4½	6	8	10	—
300-350 MCM	5	8	10	12	—
400-500 MCM	6	8	10	12	14
600-700 MCM	8	10	12	14	16
750-900 MCM	8	12	14	16	18
1,000-1,250 MCM	10	—	—	—	—
1,500-2,000 MCM	12	—	—	—	—

Bending space at terminals shall be measured in a straight line from the end of the lug or wire connector (in the direction that the wire leaves the terminal) to the wall or barrier.

WIRING METHODS AND MATERIALS

enclosure and the surface on which they are mounted (see NEC, Article 373-2).

8-83 When mounting cabinets and cutout boxes, how much set-back is allowed?

If mounted in concrete or other noncombustible material, they must not be set back more than ¼ inch. If mounted in wood or other combustible material, they must be mounted flush with the finished surface (see NEC, Article 373-3).

8-83a Is there any limitation on the deflection of conductors in cabinets or cutout boxes?

Yes, use Table 373-6a of the NEC.

8-83b Is there any limitation of the deflection of conductors at terminal connections?

Yes, use Table 373-6(a) of the NEC. If conductors are deflected or bent too sharply, damage or breakage of conductors may result. (See NEC, Section 373-6(b).)

8-84 Can unused openings in cabinets or cutout boxes be left open?

No, they must be effectively closed (see NEC, Article 373-4).

8-85 What type of bushings must be used on conduits with No. 4 wire or larger?

The bushings must be of the insulating type, or other approved methods must be used when insulating the bushings. Where bonding bushings are also required, combination bonding and insulating bushings are available (see NEC, Article 373-6).

8-86 May enclosures for switches or overcurrent devices be used as raceways or junction boxes?

Enclosures for switches or overcurrent devices shall not be used as junction boxes, auxillary gutters or raceways for conductors feeding through or tapping off to other switches or overcurrent devices, unless adequate space is provided so that the conductors do not fill the wiring space at any cross section to more than 40% of the cross-sectional area of the space, and so that the conductors, splices and taps do not fill the wiring space at any cross section to more than 75% of the cross-sectional area of the space. (See NEC, Section 373-8.)

8-87 May switches be connected in the grounded conductor?

WIRING METHODS AND MATERIALS

No, unless the ungrounded conductor is simultaneously opened at the same time as the grounded conductor (see NEC, Article 380-1).

8-87a Are signal circuits or control conductors for a motor and starter in auxiliary gutters, considered as current carrying conductors?

No. (See NEC, Section 374-5, Exception No. 2.)

8-87b How does derating affect conductors in auxiliary gutters?

When the number of current-carrying conductors are 30 or less, the correction factors specified in Note 8 with Tables 3 and 4 of Chapter 7 do not apply. (See NEC, Section 374-6.)

8-88 Can the grounded conductor be switched when using three- or four-way switches?

No, the wiring must be arranged so that all the switching is done in the ungrounded conductor. When the wiring between switches and outlets is run in metal raceways, both polarities must be in the same enclosure (see NEC, Article 380-2).

8-89 In what position must knife switches be mounted?

Single-throw knife switches must be so mounted that gravity will not tend to close them. Double-throw knife switches may be mounted either vertically or horizontally; however, if mounted vertically, a locking device must be provided to insure that the blades remain in the open position when so set (see NEC, Article 380-6).

8-90 Do switches and circuit breakers have to be accessible and grouped?

Yes, switches and circuit breakers, as far as is practicable, must be readily accessible and grouped (see NEC, Article 380-8).

8-91 How must the blades of knife switches be connected?

Unless they are the double-throw type switches, they must be connected so that when open, the blades will be "dead" (see NEC, Article 380-7).

8-92 May knife switches rated at 250 volts at more than 1,200 amperes or 600 volts at more than 600 amperes be used as disconnecting switches?

No, they may be used for isolation switches only; for interrupting currents such as these or larger, oil circuit breakers must be used (see NEC, Article 380-13).

8-93 May fuses be used in multiple on switches?

No, a fused switch shall not have fuses in parallel. (See NEC, Section 240-8.)

8-94 What is the maximum number of overcurrent devices permitted on one panelboard?

Not over 42 overcurrent devices of a lighting and appliance branch-circuit panelboard can be installed in any one cabinet or cutout box. A two-pole circuit breaker is considered as two overcurrent devices, and a three-pole circuit breaker is considered as three overcurrent devices in the interpretation of this question (see NEC, Article 384-15).

8-96 How would you define a lighting and appliance branch circuit panelboard?

For the purpose of this article, a lighting and appliance branch-circuit panelboard is one having more than 10% of its overcurrent devices rated 30 amperes or less, for which neutral connections are provided. (See NEC, Section 384-14.)

CHAPTER 9

Batteries and Rectification

9-1 What is a cell (as referred to in connection with batteries)?
A cell is a single unit capable of producing a DC voltage by converting chemical energy into electrical energy.

9-2 What is a primary cell?
A primary cell is a cell that produces electric current from an electromechanical reaction but is not capable of being recharged.

9-3 What is a secondary cell?
A cell that is capable of being recharged by passing an electric current through it in the opposite direction from the discharging current.

9-4 What is a battery?
Two or more dry cells or storage cells connected together to serve as a single DC voltage source.

9-5 What are the three most common types of cells?
The lead-acid cell, the alkaline cell, and the ordinary dry cell.

9-6 What type of cell is used as an automobile battery?
The lead-acid cell is the one in common use, although the highly expensive and highly dependable nickel-cadmium battery is used to some extent.

9-7 What is the rated voltage of a standard lead-acid cell?
2 volts.

9-8 How are automobile batteries rated?
In ampere-hours.

9-9 Explain the meaning of ampere-hours.
At full charge, the battery is capable of delivering X number of amperes for Y number of hours; e.g., a 100-ampere-hour battery would be capable of delivering 10 amperes for 10 hours, 1 ampere for 100 hours, etc.

BATTERIES AND RECTIFICATION

9-10 What is the specific gravity of the acid in a fully charged (lead-acid) car battery?
Approximately 1.280 to 1.300.

9-11 What is the specific gravity of a discharged car battery?
Approximately 1.200 to 1.215.

9-12 Is it necessary to replace the acid in a car battery?
Under normal conditions, no.

9-13 When adding water to a car battery, what precautions should be observed?
Use distilled water only, and only fill the cells to their prescribed levels.

9-14 Will a discharged car battery freeze easier than a fully charged battery?
Yes.

9-15 Why is it necessary to occasionally recharge a lead-acid battery, even though it is not being used?
A lead-acid battery not in use will gradually lose its charge, and if it is left in an uncharged condition, the material on the plates will flake off and short circuit the plates, thereby causing a shorted cell or cells.

9-16 How may a lead-acid battery be recharged?
By connecting it to a DC source at slightly higher than battery voltage and passing a high current through it in the opposite direction from the discharging current.

9-17 How can you obtain a DC source of power from an AC power source?
By means of AC power rectification.

9-18 Name some rectifiers in common use.
Vacuum-tube rectifiers and semiconductor-diode rectifiers, such as silicon and germanium.

9-19 How does a vacuum-tube rectifier operate?
A vacuum-tube rectifier converts an alternating current into a unidirectional (direct) current.

9-20 How does a contact, or barrier-layer, rectifier work?
It allows passage of current through the contact surface of two materials much more easily in one direction than in the other direction. The contact, or boundary, surface between the two materials is called the barrier, or blocking, layer.

Batteries and Rectification

9-21 Name some barrier-layer rectifiers.
Copper oxide cells, selenium rectifier cells, magnesium-copper sulfide cells, and semiconductor rectifiers, such as germanium and silicon.

9-22 Name two common rectifier circuits.
Half-wave rectifier and full-wave rectifier.

9-23 What is a half-wave rectifier?
A rectifier that only passes one-half of an AC wave.

9-24 What is a full-wave rectifier?
A rectifier that passes both halves of an AC wave.

9-25 Draw two cycles of an alternating current; show the output waveform from a half-wave rectifier and the output waveform from a full-wave rectifier. (See Fig. 1.)

9-26 Can the output wave from half- and full-wave rectification be smoothed out?
Yes, by the proper use of capacitors in the rectifier circuit.

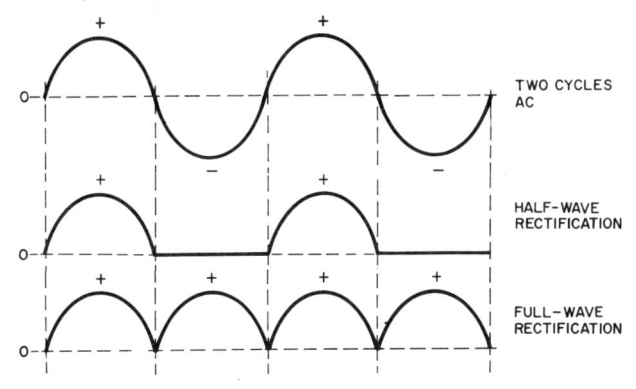

Fig. 1. Two cycles of an alternating current, with the output waveforms from a half-wave rectifier and a full-wave rectifier.

9-27 Name two mechanical methods of producing DC from AC.
A motor-generator set and a rotary converter.

9-28 What is the voltage of the ordinary dry cell?
1½ volts.

9-29 When more voltage than one cell produces is required, how should the cells be connected?
In series.

BATTERIES AND RECTIFICATION

9-30 When more current is required than one cell can supply, how should the cells be connected?
   In parallel.

9-31 How many cells are there in a 12-volt car battery, and how are they connected?
   There are six cells connected in series. Each cell supplies 2 volts, and, when 6 are connected in series, they supply the required 12 volts.

CHAPTER 10

Voltage Generation

10-1 What is a thermocouple?
A thermocouple consists of two dissimilar metals connected together at one point; when heat is applied to this point, a voltage will be generated.

10-2 Illustrate a thermocouple. (See Fig. 1.)

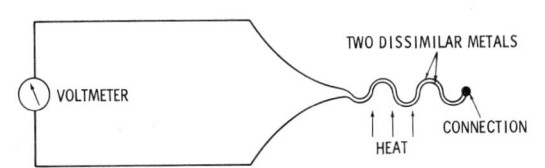

Fig. 1. A diagrammatic representation of a thermocouple.

10-3 What are some uses of a thermocouple?
As electric thermometers—the voltmeter connected across the circuit is calibrated in degrees of temperature. As a source of electricity on gas furnaces—for operating the valves on the furnace without an external source of power.

10-4 Our normal source of power comes from generators or alternators. By what principle is this power generated in these devices?
It is generated by conductors cutting magnetic lines of force, or by the lines of force cutting the conductors.

10-5 When conductors cut lines of force, what factors determine the voltage generated?
The strength of the magnetic field, the number of conductors in series, and the speed at which the field is cut.

10-6 Basically, what must a generator consist of?
A magnetic field—usually an electromagnet energized by

187

Voltage Generation

direct current, an armature or rotor—coils on an iron frame, and some device for taking the current from the rotor or armature—a commutator or slip rings.

10-7 In a standard DC generator, what type of voltage is generated?

The voltage generated is AC.

10-8 What is the main difference between an alternator and a DC generator?

The method by which the generated power is taken from the generator. The alternator uses slip rings, and the DC generator has a commutator, which takes the current from the coils in such a manner that it always flows in the same direction.

10-9 Draw a simple alternator. (See Fig. 2.)

10-10 What does the waveform taken from an alternator look like? (See Fig. 3.)

10-11 Draw a simple DC generator with only two commutator segments. (See Fig. 4.)

10-12 What does the waveform from a commutator with only two segments look like? (See Fig. 5.)

10-13 When more segments are added to the commutator on a DC generator, what happens to the waveform? Explain and illustrate.

The waveform tends to straighten out into a more nearly pure DC waveform (see Fig. 6).

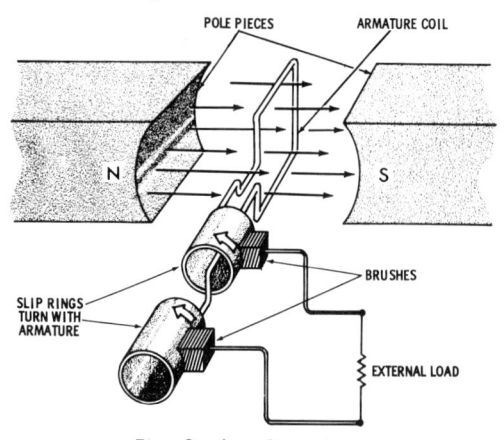

Fig. 2. An alternator.

Voltage Generation

10-14 How is the voltage output on a DC generator regulated?

It may be regulated by changing the strength of the field excitation by means of a rheostat, or by controlling the generator speed.

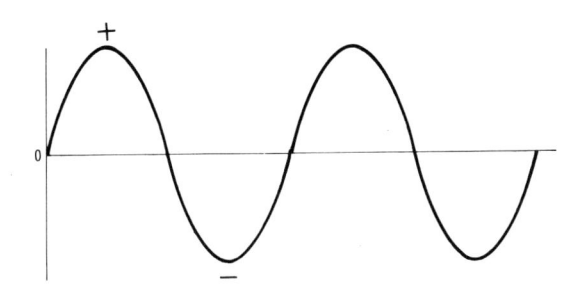

Fig. 3. The waveform from the output of an alternator.

10-15 How is the voltage output of an alternator controlled?

Since the frequency of the output must be maintained constant, the field excitation must be varied with a rheostat.

10-16 What is the formula for figuring the output frequency of an alternator?

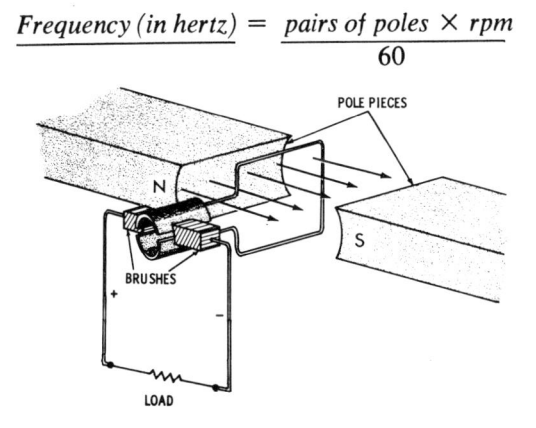

$$Frequency\ (in\ hertz) = \frac{pairs\ of\ poles \times rpm}{60}$$

Fig. 4. A DC generator with two commutator segments.

VOLTAGE GENERATION

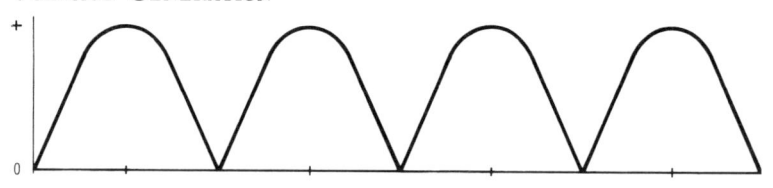

Fig. 5. The waveform from the output of a DC generator having two commutator segments.

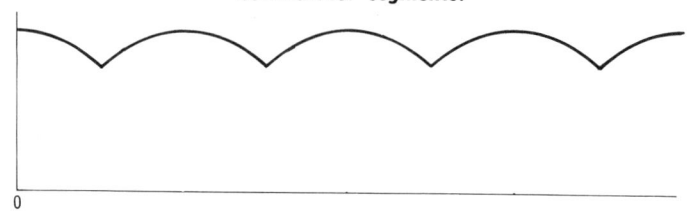

Fig. 6. The addition of commutator segments on a DC generator has the effect of straightening out the output waveform.

10-17 What is the frequency of a 4-pole alternator, driven at 1,800 rpm?

A 4-pole alternator has 2 pairs of poles. Therefore, the frequency is

$$\frac{2 \times 1,800}{60} = \frac{3,600}{60} = 60 \; hertz$$

10-18 How is the field on a DC generator usually excited?

It is usually self-excited; that is, as the generator starts generating, due to residual magnetism, part of the output is fed into the field.

10-19 How is the field on an alternator usually excited?

A DC generator is usually attached to the end of the alternator shaft; it is from this generator that the excitation current for the alternator field is received.

10-20 On an alternator, does the DC part (the poles) or the AC part (the conductors) rotate?

It is immaterial which part rotates. However, the DC field is usually made the rotating part, and the stator is usually the AC part of the device. This is because the DC field excitation can be of relatively low voltage, and it is easier to insulate a rotating part for low voltage than for high voltage. The AC output is

VOLTAGE GENERATION

usually a much higher voltage, and it is much more practical to insulate the stator for the high voltage; also, with this arrangement, there are no brushes required on the output side.

10-21 With a rotating DC field on an alternator, what is used for the DC input?

Slip rings are used.

10-22 On a DC generator, which parts, if any, must be laminated? Why?

The armature must be laminated to lower the hysteresis and eddy-current losses. It is not necessary to laminate the field, because it is strictly DC, and there would be no hysteresis or eddy-current losses.

10-23 On an alternator, what parts must be laminated? Why?

The stator must be laminated, because of hysteresis and eddy-current losses. The rotor need not be laminated because the magnetic flux is steady.

10-24 Why is it preferable to generate alternating rather than direct current?

Alternating current can be changed in voltage by means of transformers; this is necessary, because to transmit power over any distance, it must be at high voltage, due to the losses incurred when transmitting power at low voltage. Direct current cannot be changed in voltage without first changing it to AC and then raising the voltage; the operation must then be reversed at the receiving end.

10-25 Is it possible to generate power with a common three-phase squirrel-cage motor?

Yes, this is what is known as an induction generator.

10-26 How can a common three-phase squirrel-cage motor be made to generate?

It must be connected in a three-phase circuit and be driven by a prime mover at a faster-than-synchronous speed. It will not generate unless it is connected in a circuit that is already supplying current.

10-27 Will a squirrel-cage motor develop wattless or true power?

It is only capable of supplying true power; the alternator must supply the wattless-power component.

CHAPTER 11

Equipment for General Use

11-1 What may flexible cord be used for?
 Flexible cord may be used only for pendants, wiring of fixtures, connection of portable lamps and appliances, elevator cables, wiring of cranes and hoists, for connection of stationary equipment to facilitate their interchange, or to prevent the transmission of noise or vibration. Fixed or stationary appliances where the fastening means and mechanical connections are designed to permit removal for maintenance and repair, or data processing cables as permitted by Section 645-2. (See NEC, Section 400-7.)

11-2 What are the prohibited uses of flexible cord?
 Flexible cord cannot be used as a substitution for the fixed wiring of a structure; run through holes in walls, floors, or ceilings; run through doors, windows, or similar openings; for attaching to building surfaces; or where concealed behind building walls, ceilings, or floors (see NEC, Article 400-8).

11-3 Are splices or taps permitted in flexible cord?
 Flexible cord shall be used only in continuous length without splice or tap when initially installed in applications permitted by Section 400-7(a) (see 11-1). The repair of hard-service flexible cords No. 12 and larger shall be permitted if conductors are spliced in accordance with Section 110-14(b) and the complete splice retains the insulation, outer sheath properties, flexibility, and usage characteristics of the cord being spliced. (See NEC, Section 400-9.)

11-4 What types of flexible cord may be used in show windows and showcases?
 Types S, SO, SJ, SJO, ST, STO, SJT, SJTO, or AFS (see NEC, Section 400-11).

11-5 What is the minimum size conductor permitted for flexible cord?

193

Equipment for General Use

The individual conductors of a flexible cord or cable shall be not smaller than the sizes shown in Table 400-4. (See NEC for proper Table.)

11-6 What are the current-carrying capacities of various flexible cords, in amperes?

Table 1 gives the allowable current-carrying capacities for not more than three current-carrying conductors in a cord. If the number of current-carrying conductors in a cord is from four to six, the allowable current-carrying capacity of each conductor will be reduced to 80% of the values in the table.

11-7 Does the grounding conductor of a flexible cord have to be identified?

Yes, it must be identified with a green color or continuous green color with one or more yellow stripes. See NEC, Section 400-22.

11-8 What is the minimum size allowed for fixture wires?

They can be no smaller than No. 18 wire (see NEC, Section 402-6).

11-9 What is the allowable current-carrying capacity for fixture wires?

Ampacity of Fixture Wires. The ampacity of Fixture wire shall not exceed the following:

Size (AWG)	Ampacity
18	6
16	8
14	17

No conductor shall be used under such conditions that its operating temperature will exceed the temperature specified in Table 2 for the type of insulation involved.

11-10 What are the requirements, with respect to exposed live parts, on lighting fixtures?

There can be no live parts exposed, except on cleat receptacles and lampholders that are located at least 8 feet above the floor (see NEC, Article 410-3).

11-11 What are the requirements for fixtures located in damp areas?

The fixtures must be approved for the location and must be so

EQUIPMENT FOR GENERAL USE

Table 1. Ampacity of Flexible Cords and Cables

(Based on Ambient Temperature of 30°C(86°F) See Section 400-13 and Table 400-4)

Size AWG	Rubber Types TP, TS Thermoplastc Types TPT, TST	Rubber Types C, PD, E, EO, EN, S, SO, SPD, SJ, SJO, SV, SVO, SP		Types AFS, AFSJ, HPD HSI, HSJO, HS, HSO, HPN	Cotton Types CFPD* Asbestos TYPES AFC* AFPD*
		Thermoplastic Types ET, ETT, ETLB, ETP, ST, STO, SRDT, SJT, SJTO, SVT, SVTO, SPT			
		A†	B†		
27**	0.5	..	..	..	..
18	..	.7	10	10	6
17	..	..	12	..	..
16	..	10	13	15	8
15	..	..	..	17	..
14	..	15	18	20	17
12	..	20	25	30	23
10	..	25	30	35	28
8	..	35	40	..	..
6	..	45	55	..	..
4	..	60	70	..	..
2	..	80	95	..	..

* These types are used almost exclusively in fixtures where they are exposed to high temperatures and ampere ratings are assigned accordingly.

**Tinsel Cord.

† The ampacities under sub-heading A apply to 3-conductor cords and other multiconductor cords connected to utilization equipment so that only 3 conductors are current carrying. The ampacties under sub-heading B apply to 2-conductor cords and other multiconductor cords connected to utilization equipment so that only 2 conductors are current carrying.

NOTE: Ultimate insulation Temperature. In no case shall conductors be associated together in such way with respect to the kind of circuit, the wiring method used, or the number of conductors that the limiting temperature of the conductors will be exceeded.

installed that moisture cannot enter the raceways, lampholders, or other electrical parts (see NEC, Article 410-4).

Equipment for General Use

Table 2. Fixture Wire

Trade Name	Type Letter	Insulation	Thickness of Insulation AWG	Mils	Outer Covering	Max. Operating Temp.	Application Provisions
Heat Resistant Rubber-Covered Fixture Wire Solid or 7-Strand	RFH-1	Heat-Resistant Rubber	18	15	Nonmetallic Covering	75°C 167°F	Fixture wiring. Limited to 300 volts.
	RFH-2	Heat-Resistant Rubber	18-16	30	Nonmetallic Covering	75°C 167°F	Fixture wiring, and as permitted in Section 725-16.
		Heat-Resistant Latex Rubber	18-16	18			
Heat-Resistant Rubber-Covered Fixture Wire Flexible Stranding	FFH-1	Heat-Resistant Rubber	18	15	Nonmetallic Covering	75°C 167°F	Fixture wiring. Limited to 300 volts.
	FFH-2	Heat-Resistant Rubber	18-16	30	Nonmetallic Covering		
		Heat-Resistant Latex Rubber	18-16	18		75°C 167°F	Fixture wiring, and as permitted in Section 725-16.

Table 2. Fixture Wire (contd.)

Thermoplastic-Covered Fixture Wire —Solid or Stranded	TF Thermoplastic	18–16............30	None	60°C 140°F Fixture wiring, and as permitted in Section 725-16.
Thermoplastic-Covered Fixture Wire —Flexible Stranding	TFF Thermoplastic	18–16............30	None	60°C 140°F Fixture wiring, and as permitted in Section 725-16.
Heat Resistant Thermoplastic-Covered Fixture Wire—Solid or Stranded	TFN Thermoplastic	18–16............15	Nylon Jacketed	90°C 194°F Fixture wiring, and as permitted in Section 725-16.
Heat Resistant Thermoplastic-Covered Fixture Wire—Flexible Stranded	TFFN Thermoplastic	18–16............15	Nylon Jacketed	90°C 194°F Fixture wiring, and as permitted in Section 725-16.

Equipment for General Use

Table 2 (Cont.)

Trade Name	Type Letter	Insulation	AWG	Thickness of Insulation Mils	Outer Covering	Max. Operating Temp.	Application Provisions
Cotton-Covered, Heat-Resistant, Fixture Wire	CF	Impregnated Cotton	18-14	30	None	90°C 194°F	Fixture wiring. Limited to 300 volts.
Asbestos-Covered, Heat-Resistant, Fixture Wire	AF	Impregnated Asbestos	18-14	30	None	150°C 302°F	Fixture wiring. Limited to 300 volts and Indoor Dry Location.
Silicone Insulated Fixture Wire Solid or 7-Strand	SF-1	Silicone Rubber	18	15	Nonmetallic Covering	200°C 392°F	Fixture wiring. Limited to 300 volts.
	SF-2	Silicone Rubber	18-14	30	Nonmetallic Covering	200°C 392°F	Fixture wiring, and as permitted in Section 725-16.
Silicone Insulated Fixture Wire Flexible Stranding	SFF-1	Silicone Rubber	18	15	Nonmetallic Covering	150°C 302°F	Fixture wiring. Limited to 300 volts.
	SFF-2	Silicone Rubber	18-14	30	Nonmetallic Covering	150°C 302°F	Fixture wiring, and as permitted in Section 725-14.

EQUIPMENT FOR GENERAL USE

Table 2. Fixture Wire (Cont.)

Fluorinated Ethylene Propylene Fixture Wire Solid or 7 Strand	PF / PGF	Fluorinated Ethylene Propylene	18-14 18-14	None Glass Braid	200°C 392°F	Fixture wiring, and as permitted in Section 725-16.
Fluorinated Ethylene Propylene Fixture Wire Flexible Stranding	PFF / PGFF	Fluorinated Ethylene Propylene	18-14 18-14	None Glass Braid	150°C 302°F	Fixture wiring, and as permitted in Section 725-16.
Extruded Polytetrafluoroethylene Solid or 7-Strand (Nickel or Nickel Coated Copper)	PTF	Extruded Polytetrafluoroethylene	18-14	None	250°C 482°F	Fixture wire, and as permitted in Section 725-16. (Nickel or nickel-coated copper)
Extruded Polytetrafluoroethylene	PTFF	Extruded Polytetrafluoro-	18-14	None	150°C 302°F	Fixture wire, and as permitted in Section 725-16. (Silver or nickel-coated copper)

EQUIPMENT FOR GENERAL USE

Table 2 (Cont.)

Flexible Stranding (No. 26-36 AWG Silver or Nickel Coated Copper)	ethylene

11-12 What precautions must be taken with fixtures installed near combustible materials?

They must be constructed, installed, or equipped with shades or guards so that combustible materials in the vicinity of the fixture will not be subject to temperatures in excess of 90°C. (194°F.) (see NEC, Article 410-5).

11-13 Can externally wired fixtures be used in show windows?

No, with the exception of chain-supported fixtures (see NEC, Article 410-7).

11-14 Where must fixtures be installed in clothes closets?

A fixture in a clothes closet shall be installed:

(1) On the wall above the closet door, provided the clearance between the fixture and the storage area where combustible material may be stored within the closet is not less than 18 inches, or

(2) On the ceiling over an area which is unobstructed to the floor, maintaining an 18-inch clearance horizontally between the fixture and a storage area where combustible material may be stored within the closet.
Note: A flush recessed fixture equipped with a solid lens is considered to be outside the closet area.

(3) Pendants shall not be installed in clothes closets (see NEC, Section 410-8).

EQUIPMENT FOR GENERAL USE

11-15 What type of wire must be used in the connection of fixtures?

The wire must have the type of insulation that will withstand the temperatures to which the fixture will be exposed. Check the temperature rating of the wire against the temperature at which the fixture operates to obtain the type of insulation required (see NEC, Article 410-11).

11-15a May branch circuit wiring pass through an outlet box that is an integral part of an incandescent fixture?

No, unless the fixture is approved for the purpose. (See NEC, Article 410-11.)

11-16 When a fixture weighs more than 50 pounds, may it be supported from the outlet box?

It must be supported independently of the outlet box (see NEC, Article 410-16).

11-17 What is the minimum size conductor allowed for fixture wires?

No. 18 wire (see NEC, Section 410-23).

11-18 What factors must be considered when choosing fixture wires?

Operating temperature, corrosion and moisture conditions, voltage, and current-carrying capacity (see NEC, Section 410-24).

11-19 What precautions must be taken in conductors for movable parts of fixtures?

Stranded conductors must be used and must be so arranged that the weight of the fixture or movable parts will not put tension on the conductors. These measures must be taken to protect the conductors (see NEC, Section 410-26).

11-19a What types of insulation are permitted within 3 inches of a ballast within a ballast compartment?

90°C (194°F) of the following types: RHH, THHN, FEP, FEPB, SA, XHHN, AVA and THW. See Table 310-2a of the NEC for type THW (see NEC, Section 410-31).

11-20 Which wire must be connected to the screw shells of lampholders?

The identified, or white conductor (see NEC, Section 410-32).

Equipment for General Use

11-21 What is the maximum wattage permitted on a medium lamp base?

300 watts (see NEC, Section 410-53).

11-22 What is the maximum lamp wattage permitted for a mogul lamp base?

1,500 watts (see NEC, Section 410-53).

11-23 For flush or recessed fixtures, what is the maximum operating temperature permitted when installed in or near combustible materials?

90°C. or 194°F. (see NEC, Section 410-65).

11-24 If a fixture is recessed in a noncombustible material, what is the maximum temperature permitted?

A maximum temperature of 150°C. or 302°F., and the fixture must be plainly marked as approved for this type of service (see NEC, Section 410-65).

11-25 What precautions must be taken on switching of discharge-lighting systems that are rated at 1,000 volts or more?

The switch must be located in sight of the fixtures or lamps, or it may be located elsewhere, provided it is capable of being locked in the open position (see NEC, Section 410-81).

11-26 When lighting fixtures are installed near grounded surfaces, what precautions must be taken?

They should be properly grounded, or, if they are not grounded, they must be insulated from all conducting materials. They must also be located at least 8 feet vertically and 5 feet horizontally away from all grounds; this includes all plumbing fixtures, steam pipe, or any other grounded metal work or grounded surfaces (see NEC, Section 410-18).

11-27 How must fixtures be grounded?

They may be connected to metal raceways, the armor of metal-clad cable, etc. if properly installed and grounded, or a separate grounding conductor not smaller than No. 14 wire may be used (see NEC, Section 410-21).

11-28 How must immersion-type portable heaters be constructed?

They must be so constructed and installed that current-carrying parts are effectively insulated from the substance in which they are immersed (see NEC, Article 422-9).

EQUIPMENT FOR GENERAL USE

11-29 Must electric flatirons have temperature-limiting devices?

Yes (see NEC, Article 422-13).

11-30 What type of disconnection means must be provided on stationary appliances?

On appliances rated at less than 300 volt-amperes or ⅛ hp, the branch-circuit overload device may serve as the disconnecting means. For stationary appliances of greater rating, the branch-circuit switch or circuit breaker may serve as the disconnecting means, provided that it is readily accessible to the user of the appliance. For ranges, dryers, and any other cord-connected appliance, the plug at the receptacle may suffice as the disconnecting means (see NEC, Section 422-23).

11-31 On motor-driven appliances, what disconnecting means must be provided, and how must it be located?

The switch or circuit breaker that serves as the disconnecting means on a stationary motor-driven appliance must be located within sight of the motor controller or be capable of being locked in the open position (see, Section 422-26).

11-32 On all space-heating systems, what is the main requirement?

All heating equipment shall be installed in an approved manner. (See NEC, Article 424-9.)

11-36 What are the requirements for heating cables?

Heating cables must be supplied complete with factory-assembled nonheating leads of at least 7 feet in length, and these leads must be of approved conductors for the purpose (see NEC, Article 424-34).

11-37 What markings must be present on heating cables?

Each unit length must have a permanent marking located within 3 inches of the terminal end of the nonheating leads, with the manufacturer's name or identification symbol, catalog number, and the rating in volts and watts or amperes. The leads on a 230-volt cable must be red; on a 115-volt cable the leads are yellow; 208-volt cable leads are blue; and on a 277-volt cable the leads are brown. (See NEC, Article 424-35.)

11-38 Must controllers or disconnecting means open all ungrounded conductors?

EQUIPMENT FOR GENERAL USE

Devices that have an "off" position must open all ungrounded conductors; thermostats without an "on" or "off" position do not have to open all ungrounded conductors (see NEC, Section 424-20).

11-39 What clearance must be given to wiring in ceilings?

Wiring above heated ceilings and contained within thermal insulation must be spaced not less than 2 inches above the heated ceiling and will be considered as operating at 50°C. Wiring above heated ceilings and located over thermal insulation having a minimum thickness of 2 inches requires no correction for temperature. Wiring located within a joist space having no thermal insulation must be spaced not less than 2 inches above the heated ceiling and will be considered as operating at 50°C. (See NEC, Article 424-36.)

11-40 What clearance is required for wiring in walls where electrical heating is used?

Wires on exterior walls must be located outside the thermal insulation. On interior walls or partitions, wiring must be located away from the heated surfaces and will be considered as operating at 40°C. (See NEC, Article 424-37.)

11-41 What are the restricted areas for heating?

Heating panels must not extend beyond the room in which they originate; cables must not be installed in closets, over cabinets that extend to the ceiling, under walls or partitions or over walls or partitions that extend to the ceiling (see NEC, Article 424-38).

11-42 If heating is required in closets for humidity control, may it be used?

Yes; low-temperature heating sources may be used (see NEC, Article 424-38).

11-43 What clearance must be provided for heating cables and panels from fixtures, boxes, and openings?

Panels and heating cables must be separated by a distance of at least 8 inches from lighting fixtures, outlets, and junction boxes, and 2 inches from ventilating openings and other such openings in room surfaces, or at least a sufficient area must be provided (see NEC, Article 424-39).

11-44 May embedded cables be spliced?

Only when absolutely necessary, and then only by approved

EQUIPMENT FOR GENERAL USE

means; in no case may the length of the cable be altered (see NEC, Section 424-40).

11-45 May heating cable be installed in walls?
No (see NEC, Article 424-41).

11-46 What is the spacing requirement on heating cable in dry wall and plaster?
Adjacent runs of heating cable not exceeding 2¾ watts per foot must be installed not less than 1½ inches on centers (see NEC, Article 424-41).

11-47 How is heating cable installed in dry board and plaster?
Heating cables may be applied only to gypsum board, plaster lath, or other similar fire-resistant materials. On metal lath or other conducting surfaces, a coat of plaster must be applied first. The entire ceiling must have a coating of ½ inch or more of thermally noninsulating sand plaster or other approved material. Cable must be secured at intervals not exceeding 16 inches with tape, staples, or other approved devices; staples cannot be used with metal lath. On dry-board installations, the entire ceiling must be covered with gypsum board not exceeding ½ inch in thickness, and the void between the upper and lower layers of gypsum board must be filled with thermally conducting plaster or other approved material (see NEC, Article 424-41).

11-48 Can the excess of the nonheating leads on heating cable be cut off?
No; they must be secured to the under side of the ceiling and embedded in the plaster. The ends have a color coding on them, and these must be visible in the junction box (see NEC, Article 424-43).

11-49 How is heating cable installed in concrete?
Panels or heating units shall not exceed 33 watts per square foot of heating area or 16½ watts per linear foot of cable. Adjacent runs of cable shall be a minimum of 1 inch apart. The cable shall be secured with frames or spreaders (nonmetallic) while the concrete is being applied. A minimum space of 1 inch must be maintained between the cable and other metallic objects. The leads extending from the concrete shall be protected by rigid metal conduit or EMT. Bushings shall be used where cable leads emerge from the floor slab. (See NEC, Article 424-44.)

EQUIPMENT FOR GENERAL USE

11-50 What tests must be run on cables during and after installation?

They must be tested for continuity and insulation resistance (see NEC, Article 424-45).

11-51 What percentages of branch-circuit ratings are allowable for use with air-conditioning units?

When supplying only the air-conditioning unit, not more than 80% of the circuit rating may be used. When lighting or other appliances are also supplied on the same branch circuit, not more than 50% of the circuit rating may be used for the air-conditioning unit (see NEC, Section 440-62).

11-52 Can a plug and receptacle be used as the disconnecting means for an air-conditioning unit?

An attachment plug and receptacle may serve as the disconnecting means for a single-phase room air-conditioning unit, rated 250 volts or less when:

(1) The manual controls on the air-conditioning units are readily accessible and located within 6 feet of the floor;
(2) An approved manually operated switch is installed in a readily accessible location within sight of the air-conditioning unit. (see NEC, Section 440-63).

11-53 What is a sealed (hermetic type) motor compressor?

This is a mechanical compressor and a motor, enclosed in the same housing, with neither an external shaft nor seal; the motor operates in the refrigerant atmosphere (see NEC, Section 440-17).

11-54 What does "in sight from" mean as it is applied to motors?

When "in sight from" is used with reference to some equipment in relation to other equipment, the term means that the equipment must be visible and within 50 feet of each other (see NEC, Article 430-4).

11-55 Draw a motor-feeder and branch-circuit diagram, showing at least 5 parts. (See Fig. 1.)

11-56 What is the purpose of a locked-rotor table?

A locked-rotor table indicates the current, when full voltage is applied to a motor, with the rotor held in a locked position. Table 3 is a typical locked-rotor table.

Equipment for General Use

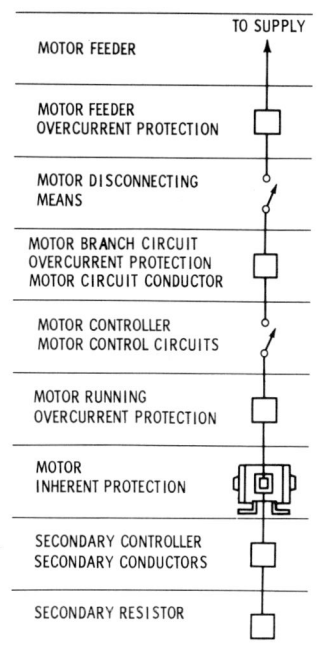

Fig. 1. A motor feeder and branch circuit.

```
                                              TO SUPPLY
         MOTOR FEEDER                              ↑
    ─────────────────────────────────────
         MOTOR FEEDER
         OVERCURRENT PROTECTION            □
    ─────────────────────────────────────
         MOTOR DISCONNECTING
         MEANS                             ⌐
    ─────────────────────────────────────
         MOTOR BRANCH CIRCUIT
         OVERCURRENT PROTECTION
         MOTOR CIRCUIT CONDUCTOR           □
    ─────────────────────────────────────
         MOTOR CONTROLLER
         MOTOR CONTROL CIRCUITS            ⌐
    ─────────────────────────────────────
         MOTOR RUNNING
         OVERCURRENT PROTECTION            □
    ─────────────────────────────────────
         MOTOR
         INHERENT PROTECTION              (M)
    ─────────────────────────────────────
         SECONDARY CONTROLLER
         SECONDARY CONDUCTORS              □
    ─────────────────────────────────────
         SECONDARY RESISTOR                □
```

11-57 How must hermetic-type refrigeration compressors be marked?

They must have a nameplate, giving the manufacturer's name, the phase, voltage, frequency, and full-load current in amperes of the motor current when the compressor is delivering the rated output. If the motor has a protective device, the nameplate must be marked "Thermal Protection". For complete details, see Section 440-3 of the NEC.

11-58 What markings must be provided on motor controllers?

Maker's name or identification, the voltage, and the current or horsepower rating (see NEC, Article 430-8).

11-59 Must the controller be the exact size for the motor it is to be used on?

No, this is not necessary; however, it must be at least as large as is necessary for the job it has to do. It may be larger, but the overloads that might be used in conjunction with it should be the proper size for the application.

EQUIPMENT FOR GENERAL USE

Table 3. Locked Rotor Indicating Code Letters

Code Letter	Kilovolt-Amperes per Horsepower with Locked Rotor
A	0 — 3.14
B	3.15 — 3.54
C	3.55 — 3.99
D	4.0 — 4.49
E	4.5 — 4.99
F	5.0 — 5.59
G	5.6 — 6.29
H	6.3 — 7.09
J	7.1 — 7.99
K	8.0 — 8.99
L	9.0 — 9.99
M	10.0 — 11.19
N	11.2 — 12.49
P	12.5 — 13.99
R	14.0 — 15.99
S	16.0 — 17.99
T	18.0 — 19.99
U	20.0 — 22.39
V	22.4 — and up

The above table is an adopted standard of the National Electrical Manufacturers Association (N.E.M.A.). The code letter indicating motor input with locked rotor must be in an individual block on the nameplate, properly designated. This code letter is to be used for determining branch-circuit overcurrent protection by reference to the NEC, Table 430-152, as provided in Section 430-52 of the NEC.

11-60 When using switches, fuses, and other disconnecting means, must the voltage of the switch coincide with that of the motor?

No. The voltage rating of the switch cannot be less than that of the application for which it is used; however, the voltage rating may be higher than the application for which it is used. This also applies to the amperage rating.

11-61 If a switch is used that has a current rating larger than is required, may it be adapted for less current?

Yes, adapters of the approved type are available for reducing the fuse size of the switch; fuses larger than the switch rating, however, cannot be used.

11-62 Can the enclosures for controllers and disconnecting means for motors be used as junction boxes?

EQUIPMENT FOR GENERAL USE

Enclosures for controllers and disconnecting means for motors shall not be used as junction boxes, auxillary gutters, or raceways for conductors feeding through or tapping off to the other apparatus unless designs are employed which provide adequate space for this purpose. See NEC, Section 373-8. See 8-86 of this book.

11-63 If a branch circuit supplies only one motor, how large must the conductor size of the branch circuit be, in comparison to the current rating of the motor?

The branch-circuit conductors should be figured at not less than 125% of the full-load current rating of the motor. If a wire size is close to but is still under the proper current rating, the next larger size conductor must be used (see NEC, Article 430-22).

11-64 What size wire must be used to connect the wound-rotor secondary of a motor?

For continuous duty, the conductor size must not be less than 125% of the full-load secondary current of the motor (see NEC, Article 430-23).

11-65 When conductors supply several motors, how can you determine the current-carrying capacity of the conductors?

Conductors cannot use less than 125% of the full-load current rating of the highest rated motor in the group and not less than the full-load current rating of the rest of the motors. The total current is the basis for calculating the wire size (see NEC, Article 430-24).

11-66 What overload-current protection must motors of more than 1 hp have?

Motors having a temperature rise of not more than 40°C. must trip at not more than 125% of full-load current (see NEC, Article 430-32).

11-67 How many conductors must be opened by motor overcurrent devices?

A sufficient number of ungrounded conductors must be opened to stop the motor (see NEC, Article 430-38).

11-68 What are the number and location of overcurrent-protection units for different motors?

See Table 4.

209

Equipment for General Use

11-69 When are three overcurrent-protection devices required on a three-phase motor?
One in each phase unless protected by other approved means. This would indicate that you should check the authority having jurisdiction.

11-70 Where the motor and driven machinery are not "in sight from" the controller location, what conditions must be met?
The controller disconnection means must be capable of being locked in the open position, or there must be a manually operated switch that will disconnect the motor from its source of supply, placed within sight of the motor location (see NEC, Article 430-86).

Table 4. Running Overcurrent Units

Kind of Motor	Supply System	Number and location of overcurrent units, such as trip coils, relays or thermal cutouts
1-phase A.C. or D.C.	2-wire, 1-phase A.C. or D.C. ungrounded	1 in either conductor
1-phase A.C. or D.C.	2-wire, 1-phase A.C. or D.C., one conductor grounded	1 in ungrounded conductor
1-phase A.C. or D.C.	3-wire, 1-phase A.C. or D.C., grounded-neutral	1 in either ungrounded conductor
2-phase A.C.	3-wire, 2-phase A.C., ungrounded	2, one in each phase
2-phase A.C.	3-wire, 2-phase A.C., one conductor grounded	2 in ungrounded conductors
2-phase A.C.	4-wire, 2-phase A.C. grounded or ungrounded	2, one per phase in ungrounded conductors
2-phase A.C.	5-wire, 2-phase, A.C. grounded neutral or ungrounded	2, one per phase in any ungrounded phase wire
3-phase A.C.	Any 3-phase	*3, one in each phase

* Exception: Unless protected by other means.

11-71 How many motors may be served from one controller?
Each motor must have its own controller; however, where the motors are rated at 600 volts or less and drive several parts of one machine, are under the protection of one overcurrent-protection device, or where a group of motors is located in one

EQUIPMENT FOR GENERAL USE

room "in sight from" the controller, a single controller may serve a group of motors (see NEC, Article 430-87).

11-72 Must the disconnecting means be marked to indicate the position it is in?

Yes, it must be plainly marked to indicate whether it is in the open or closed position (see NEC, Article 430-104).

11-73 May the service switch be used as the disconnecting means?

Yes, if it is "in sight from" the controller location, or if it is capable of being locked in the open position (see NEC, Article 430-106).

11-74 Must the disconnecting means be readily accessible?

Yes (see NEC, Article 430-107).

Table 5. Full-Load Currents in Amperes
Single-Phase Alternating-Current Motors

The following values of full-load currents are for motors running at usual speeds and motors with normal torque characteristics. Motors built for especially low speeds or high torques may have higher full-load currents, and multispeed motors will have full-load current varying with speed, in which case the nameplate current ratings shall be used.

To obtain full-load currents of 208- and 200-volt motors, increase corresponding 230-volt motor full-load currents by 10 and 15 percent, respectively.

The voltages listed are rated motor voltages. Corresponding nominal system voltages are 110 to 120 and 220 to 240.

HP	115V	230V
1/6	4.4	2.2
1/4	5.8	2.9
1/3	7.2	3.6
1/2	9.8	4.9
3/4	13.8	6.9
1	16	8
1 1/2	20	10
2	24	12
3	34	17
5	56	28
7 1/2	80	40
10	100	50

11-75 What should be the current-carrying capacity of the disconnecting means?

It must have a current-carrying capacity of at least 115% of the nameplate current rating of the motor (see NEC, Article 430-110).

Equipment for General Use

11-77 A single-phase motor draws 32 amperes under full-load condition. How large must the fuses and branch-circuit switch be?

They must be rated at 300% of the full-load current, or about 100 amperes each.

Table 6. Full-Load Current
Two-Phase Alternating-Current Motors (4-wire)

The following values of full-load current are for motors running at speeds usual for belted motors and motors with normal torque characteristics. Motors built for especially low speeds or high torques may require more running current, and multi-speed motors will have full-load current varying with speed, in which case the nameplate current rating shall be used. Current in common conductor of 2-phase, 3-wire system will be 1.41 times value given.

The voltages listed are rated motor voltages. Corresponding nominal system voltages are 110 to 120, 220 to 240, 440 to 480 and 550 to 600 volts.

HP	Induction Type Squirrel-Cage and Wound Rotor Amperes				Synchronous Type †Unity Power Factor Amperes				
	115V	230V	460V	575V	2300V	220V	440V	550V	2300V
½	4	2	1	.8					
¾	4.8	2.4	1.2	1.0					
1	6.4	3.2	1.6	1.3					
1½	9	4.5	2.3	1.8					
2	11.8	5.9	3	2.4					
3		8.3	4.2	3.3					
5		13.2	6.6	5.3					
7½		19	9	8					
10		24	12	10					
15		36	18	14					
20		47	23	19					
25		59	29	24		47	24	19	
30		69	35	28		56	29	23	
40		90	45	36		75	37	31	
50		113	56	45		94	47	38	
60		133	67	53	14	111	56	44	11
75		166	83	66	18	140	70	57	13
100		218	109	87	23	182	93	74	17
125		270	135	108	28	228	114	93	22
150		312	156	125	32		137	110	26
200		416	208	167	43		182	145	35

† For 90 and 80 percent power factor the above figures should be multiplied by 1.1 and 1.25 respectively.

EQUIPMENT FOR GENERAL USE

11-78 In a group of four motors, one motor draws 10 amperes, one draws 45 amperes, and two draw 75 amperes. What size conductors must be used for the feeder circuit?

$$75 \text{ amperes} \times 125\% = 94 \text{ amperes}$$
$$94 + 75 + 45 + 10 = 224 \text{ amperes}$$

The feeder conductors, therefore, must be able to handle 224 amperes. The size of the conductors will be 4/0 for Type RH, RUH, RH-RW, RHW, THW, or THWN cable, or 300,000 CM for Type R, RW, RU, RUW, T, or TW cable.

Table 7. Full-Load Current*

Three-Phase Alternating-Current Motors

	Induction Type Squirrel-Cage and Wound Rotor Amperes				Synchronous Type †Unity Power Factor Amperes				
HP	115V	230V	460V	575V	2300V	220V	440V	550V	2300V
½	4	2	1	.8					
¾	5.6	2.8	1.4	1.1					
1	7.2	3.6	1.8	1.4					
1½	10.4	5.2	2.6	2.1					
2	13.6	6.8	3.4	2.7					
3		9.6	4.8	3.9					
5		15.2	7.6	6.1					
7½		22	11	9					
10		28	14	11					
15		42	21	17					
20		54	27	22					
25		68	34	27		54	27	22	
30		80	40	32		65	33	26	
40		104	52	41		86	43	35	
50		130	65	52		108	54	44	
60		154	77	62	16	128	64	51	12
75		192	96	77	20	161	81	65	15
100		248	124	99	26	211	106	85	20
125		312	156	125	31	264	132	106	25
150		360	180	144	37		158	127	30
200		480	240	192	49		210	168	40

For full-load currents of 208- and 200-volt motors, increase the corresponding 230-volt motor full-load current by 10 and 15 percent, respectively.

213

Equipment for General Use

*These values of full-load current are for motors running at speeds usual for belted motors and motors with normal torque characteristics. Motors built for especially low speeds or high torques may require more running current, and multi-speed motors will have full-load current varying with speed, in which case the nameplate current rating shall be used.

†For 90 and 80 percent power factor the above figures shall be multiplied by 1.1 and 1.25 respectively.

The voltages listed are rated motor voltages. Corresponding nominal system voltages are 110 to 120, 220 to 240, 440 to 480 and 550 to 600 volts.

Since sealed (Hermetic-Type) motor compressors usually have different nameplate markings, and require special consideration, it is well to separate them from the conventional types of motors, as covered in Article 430.

Most sealed (Hermetic-Type) motor compressors are marked in amperes instead of horsepower. These amperes in most cases must be converted into horsepower, using Tables 5, 6, 7.

11-79 Air-conditioning and refrigeration equipment was separated from **Motors** in the 1971 NEC under Article 430. It now appears as a new article in the code as Article 440.

CHAPTER 12

Motors

12-1 What is a DC motor?

This is a motor designed and intended for use with direct current only.

12-2 If a DC motor is connected across an AC source, what will be the results? Why?

Direct-current flow is only obstructed by resistance, while alternating current is obstructed by both resistance and inductive reactance. Therefore, when a DC motor is connected across an AC source, the current on AC will be much less than that of DC. The motor would run, however, but it would not carry the same load as it would on DC. There would be more sparking at the brushes. The armature is made up of laminations, but the field is not. The eddy currents in the field would therefore cause the motor to heat up and eventually burn on AC; this would not happen on DC.

12-3 What is the difference between a DC motor and a DC generator?

Fundamentally there is none. A DC motor will generate electricity if driven by some prime mover, and a DC generator will act as a motor if connected across a DC source.

12-4 What are the three fundamental types of DC motors?

They are the series-wound motors, the shunt-wound motors, and the compound-wound motors.

12-5 What is a series-wound motor? Draw a schematic of one.

In a series-wound motor, the field is wound in series with the armature (see Fig. 1).

MOTORS

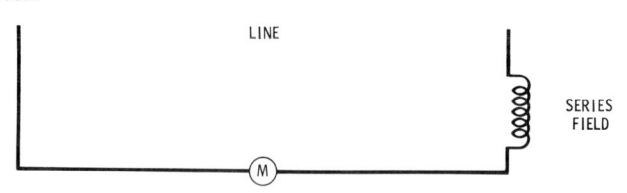

Fig. 1. A series-wound motor.

12-6 What is a shunt-wound motor? Draw a schematic of one.

A shunt-wound motor has the field winding in shunt, or parallel, with the armature. The field consists of many turns of small wire, since it must be able to handle the line voltage that is connected across it (see Fig. 2).

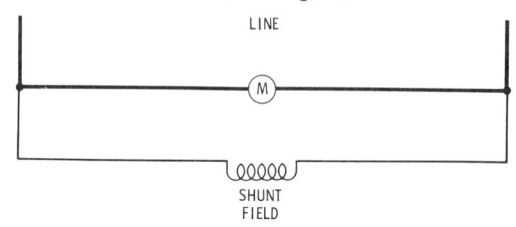

Fig. 2. A shunt-wound motor.

12-7 What is a compound-wound motor? Draw a schematic of one.

A compound-wound motor is a combination of a series- and a shunt-wound motor and has better speed regulation than either one (see Fig. 3).

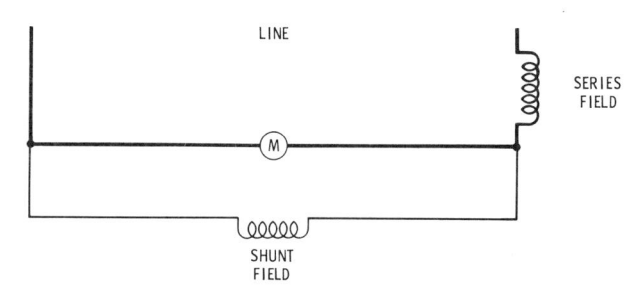

Fig. 3. A compound-wound motor.

12-8 If the field on a DC motor were opened, what would happen?

The motor would try to run away with itself, or, in other words, the motor would reach a very high speed and might destroy itself.

12-9 What are the advantages and disadvantages of DC motors as compared to AC motors?

Speed control of DC motors is much easier, making them more versatile for use where a wide range of speeds is required. They may be used for dynamic braking; that is, a motor on an electric train will act as a motor when required, but when going down hill, it can be used as a generator, thereby putting current back into the line; in generating, it requires power, so it acts as a brake. DC motors, however, require more maintenance than most AC motors. In order to use a DC motor, special provisions must be made.

12-10 What is a motor-generator set?

This is a generator driven by a motor, with both devices mounted on a common base. It might be an AC motor driving a DC generator or a DC motor driving an alternator.

12-11 What is a rotary converter?

This is a self-contained unit having two or more armature windings and a common set of field poles. One armature winding receives the direct current and rotates, thus acting like a motor, while the others generate the required voltage, and thus act as generators.

12-12 If a DC motor is rapidly disconnected from the line, what must be provided?

A means of shorting out the motor, either directly or through a resistance, so that the collapsing magnetic field does not induce a high voltage.

12-13 What different types of alternating current are used on motors?

Single-phase, two-phase, three-phase, and six-phase; 25-cycle, 50-cycle, 60-cycle, and other frequencies may also be used.

12-14 May single-phase motors be run on two- or three-phase lines?

Yes, if they are connected to only two phase wires.

MOTORS

12-15 Can a three-phase motor be run on a single-phase line?

Yes, but a phase splitter must be used.

12-16 What is a phase splitter?

This is a device, usually composed of a number of capacitors connected in the motor circuit, that produces, from a single input wave, two or more output waves which differ in phase from each other.

12-17 When using a phase splitter, there will be some strange current readings noted. What will they be?

At no-load conditions, the current on the three motor leads will be unbalanced; one will have a high current and the other two will have a low current. As the motor is loaded, these currents will begin to balance out, and, at full-load conditions, they will have equalized.

12-18 Why would you use a phase splitter and a three-phase motor instead of a single-phase motor?

It is possible that at the moment single-phase power is all that is available, but in the future three-phase power is expected. Therefore, if you purchase the three-phase motor and a phase splitter, the wiring will be in, the motor will be at the desired location, and the expenses will be cut down. There is also less maintenance on three-phase motors; this one fact will often influence the use of a phase splitter.

12-19 What causes a motor to turn?

There are two parts to a motor, a stator, or field, and a rotor, or armature. Around one part there exists a magnetic field from the line current, and in the other part there is an induced current which causes a magnetic field of opposite polarity. These mag-

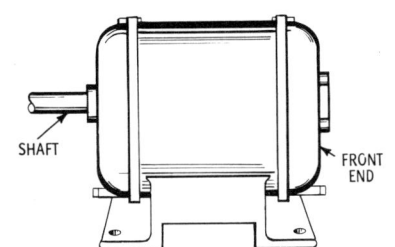

Fig. 4. The "front end" of a motor.

MOTORS

netic fields repel one another, thereby causing the rotating member to turn.

12-20 What is termed the "front end" of a motor? Draw a sketch.

The end opposite the shaft is termed the "front end" (see Fig. 4).

12-21 What is the standard direction of rotation for a motor?

Counterclockwise; that is, the shaft of the motor appears to be turning counterclockwise when looking at the front end of the motor.

12-22 Illustrate the standard and opposite directions of rotation of a motor. (See Fig. 5.)

12-23 What are several types of three-phase motors?

Squirrel-cage, wound-rotor, synchronous.

12-24 What is a squirrel-cage motor?

This is a motor with the winding on the stator, or line winding. The rotor consists of a winding made of bars that are permanently short circuited at both ends by a ring.

12-25 How are the voltage and current produced in a squirrel-cage rotor?

The stator can be considered the primary winding and the rotor the short-circuited secondary winding of a transformer. Thus, the voltage and current are induced in the rotor.

12-26 How is regulation obtained in a motor?

The motor at start is similar to a transformer with a shorted secondary. The current in the rotor and stator will be high. As the motor approaches its rated speed, the rotor induces a voltage into the stator in opposition to the line voltage; this is

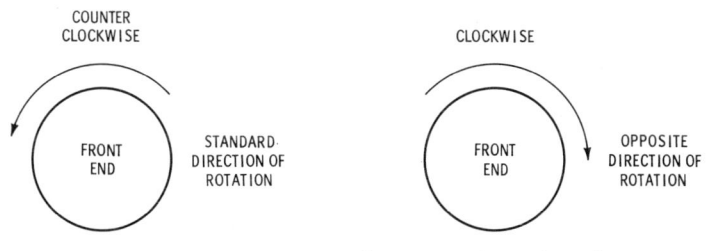

Fig. 5. The standard and opposite directions of rotation of a motor.

MOTORS

called *counter emf*. The line current is then reduced in proportion to the speed.

12-27 Does an AC motor (other than a synchronous motor) run at synchronous speed?

No. It must slip below synchronous speed so that there will be an effective voltage produced.

12-28 What is a synchronous motor?

A synchronous motor is almost exactly the same as an alternator. The field must be excited by DC. The motor runs at the same speed or at a fixed multiple of the speed of the alternator supplying the current for its operation. Should it slip, the motor will pull out and stop, since it must run pole for pole with the alternator.

12-29 How do synchronous motors differ from alternators?

They may be just like alternators; however, if they are, they will not be self-starting and will have to be started by some means until they approach synchronous speed, at which time they can be connected to the line and pull into speed. Most synchronous motors have a squirrel-cage winding in addition to the DC field. They start as a squirrel-cage motor, and, when they are about up to the speed of the alternator, the DC field is energized; the poles then lock in position with the revolving field of the armature, and the rotor revolves in synchronization with the supply circuit.

12-30 If the field of a synchronous motor is underexcited, what will happen to its power factor?

It will lag.

12-31 If the field of a synchronous motor is overexcited, what will happen to the power factor?

It will lead.

12-32 What is a synchronous capacitor?

It is a synchronous motor running without mechanical load on the line, with its field overexcited for power-factor correction.

12-33 In what direction does the rotor turn in an alternating-current motor?

It always turns in the direction of the rotating field (see Fig. 6).

MOTORS

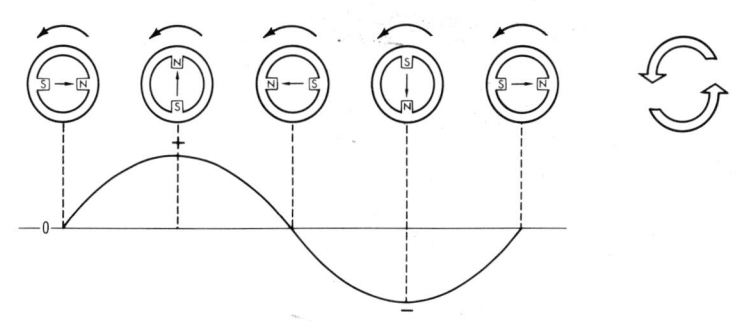

Fig. 6. The direction of rotor rotation in an AC motor.

12-34 On three-phase motors, how may the fields, or stator stator windings, be connected internally?

They may be connected in either a Delta or a Wye arrangement.

12-35 Draw a schematic of a squirrel-cage motor connected in a Delta arrangement. (See Fig. 7.)

12-36 Draw a diagram of a Delta-wound motor; number the coils, and show how they must be connected for a motor that can be used on a 240/480-volt system. (See Fig. 8.)

12-37 How are motor leads numbered?

T_1, T_2, T_3, T_4, etc.

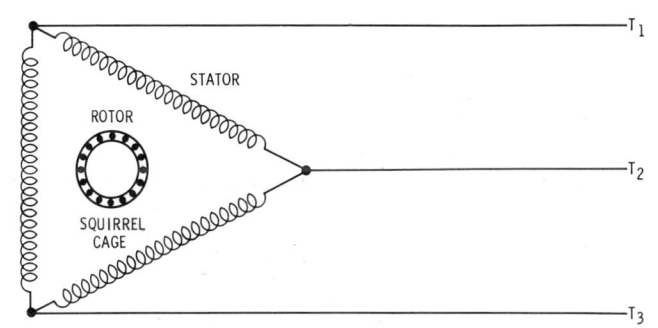

Fig. 7. A squirrel-cage motor connected in a Delta arrangement.

12-38 Draw a diagram of a Wye-wound motor; number the coils and show how they must be connected for a motor that can be used on 240/480 volts. (See Fig. 9.)

MOTORS

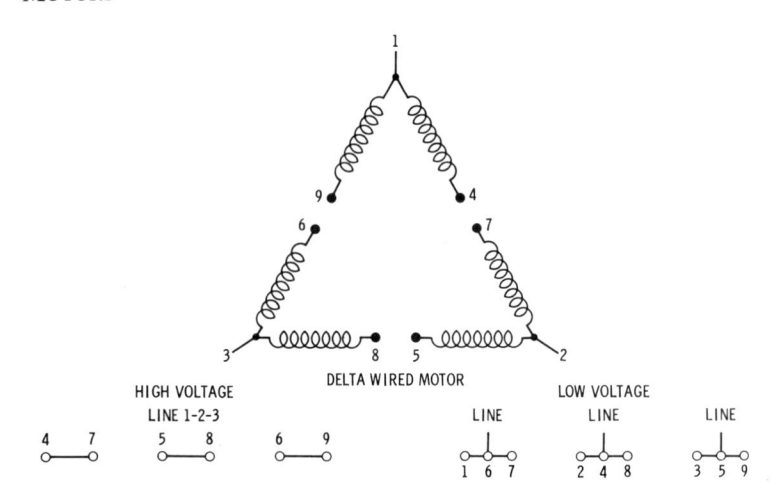

Fig. 8. A Delta-wound motor for use on 240/480 volts.

12-39 What is an easy method of remembering which leads to connect together for a Wye-wound motor that has two voltage ratings?

First, draw the windings in a Wye arrangement. Then, starting at one point, draw a spiral so that it connects in order with all six coils. Begin at the starting point and number from 1 to 9 as you progress around the spiral (see Fig. 10).

12-40 What are the different kinds of single-phase motors?

Split-phase, capacitor, capacitor-start, capacitor-start-capacitor-run, shading-pole, repulsion, repulsion-induction, and repulsion-start-induction-run.

12-41 What is a wound-rotor three-phase motor?

This is a three-phase motor that has another three-phase winding instead of a squirrel-cage rotor, the terminals of which are connected to three slip rings. Brushes ride these slip rings and deliver the current to an external three-phase rheostat or variable resistor. At start, all the resistance is in the circuit; as the motor picks up speed, the resistance is gradually decreased until finally the slip rings are short circuited.

12-42 Can a wound-rotor motor be used as a variable-speed motor?

Yes, if it has been properly designed for this purpose.

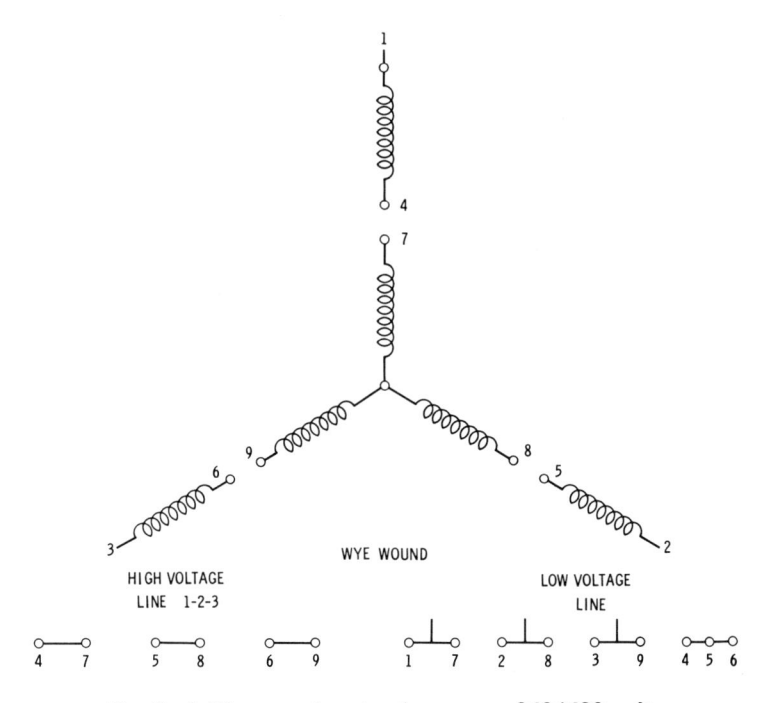

Fig. 9. A Wye-wound motor for use on 240/480 volts.

12-43 How can you reverse the direction of a three-phase squirrel-cage motor?

By transposing any two of the motor leads.

12-44 How can you change the rotation of a wound-rotor motor?

By transposing any two of the line leads.

12-45 Can a wound-rotor motor be reversed by transposing any two leads from the slip rings?

No. There is only one way to reverse its direction and that is by transposing any two line leads.

12-46 Draw a schematic diagram of a wound-rotor motor. (See Fig. 11.)

12-47 What is a split-phase motor?

This is a single-phase induction motor equipped with an auxiliary winding connected in shunt with the main stator winding

Motors

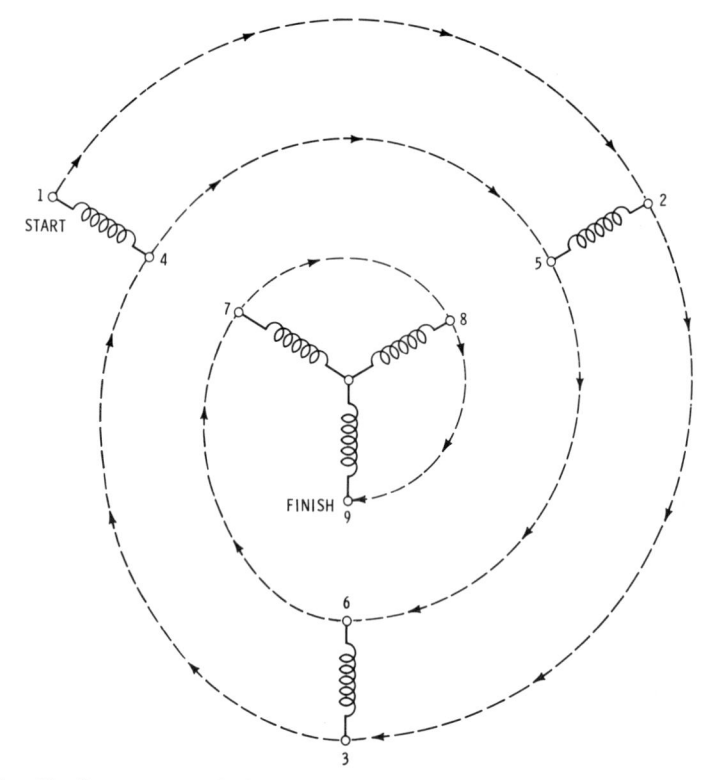

Fig. 10. One easy method of remembering the proper connections for a Wye-wound motor with two voltage ratings.

and differs from it in both phase and spacing. The auxiliary winding is usually opened by a centrifugal device when the motor has reached a predetermined speed. The fields of these two windings act on the rotor to produce a small starting torque. Once the motor is started, it produces its own rotating field, and it no longer requires the starting torque.

12-48 **Does a split-phase motor have a high or low starting torque?**

It has a low starting torque.

12-49 **How can you reverse the rotation of a split-phase motor?**

By reversing the leads to the running or the starting winding, but not both.

224

MOTORS

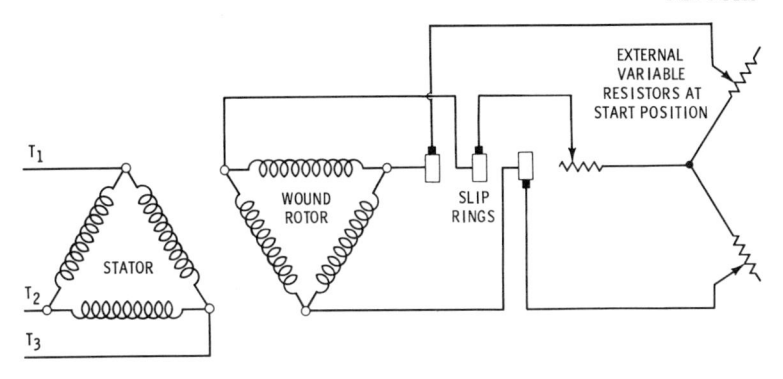

Fig. 11. The schematic representation of a wound-rotor motor.

12-50 Draw two schematic diagrams illustrating the reversal of a split-phase motor by reversing the running winding leads. (See Fig. 12.)

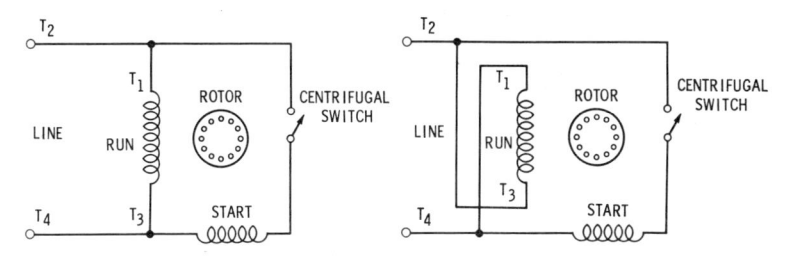

Fig. 12. The reversal of a split-phase motor is accomplished by reversing the running winding leads.

12-51 What is a capacitor-start motor?

Fundamentally this motor is very similar to the split-phase motor, except that the starting winding has a few more turns and consists of heavier wire than the starting winding of a split-phase motor; there is also a large electrolytic capacitor connected in series with the starting winding. The capacitor and starting winding are cut out of the circuit as soon as the motor reaches approximately 70% of its full speed.

12-52 Does the capacitor-start motor have a high or low starting torque?

It has a high starting torque. For this reason and the fact that it requires little maintenance, it is fast replacing the re-

Motors

pulsion types of motors; it is also cheaper to build, and the rewinding is more practical than the repulsion-type motors.

12-53 How can you reverse the direction of rotation of a capacitor-start motor?

By reversing either the runing- or starting-coil leads where they are connected to the line; do not reverse both.

12-54 Are capacitor-start motors usually dual-voltage or single-voltage motors?

They are usually dual-voltage motors.

12-55 If a capacitor-start motor is rated as a 115/230-volt motor (the capacitor is usually a 115-volt capacitor), how is it connected, when the motor is used on 230 volts, to protect the capacitor from being damaged?

The running winding is in two sections so that it may be connected in parallel for 115 volts and connected in series for 230 volts. The capacitor is in series with the starting winding; therefore, when operating at 115 volts, the winding leads are connected across the line at start, and when operating at 230 volts, one lead is connected to one side of the line, and the other lead is connected to the midpoint of the running windings.

12-56 Draw two schematic diagrams of a capacitor-start motor, with one showing the 115-volt connection and one showing the 230-volt connection. (See Fig. 13.)

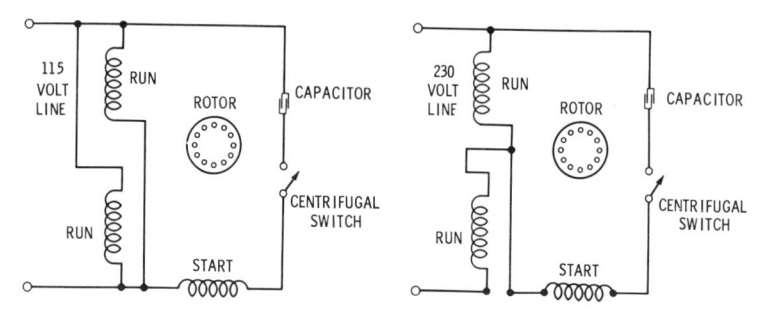

Fig. 13. A capacitor-start motor, first connected across 115 volts and then across 230 volts.

12-57 What is a capacitor-start-capacitor-run motor?

This is the same as the capacitor-start motor, except that it has an extra capacitor (oil type) connected in the starting wind-

MOTORS

ing, which is always in the motor circuit. The starting capacitor is only in the starting circuit when the motor is started and is disconnected by means of a centrifugal switch.

12-58 What are the advantages of a capacitor-start-capacitor-run motor over the capacitor-start motor?

It is smoother running, has a higher power factor, and consequently draws less current than the capacitor-start motor.

12-59 How can the direction of a capacitor-start-capacitor-run motor be changed?

By reversing the leads to either the starting winding or the running winding, but not to both.

12-60 Draw a schematic diagram of a capacitor-start-capacitor-run motor. (See Fig. 14.)

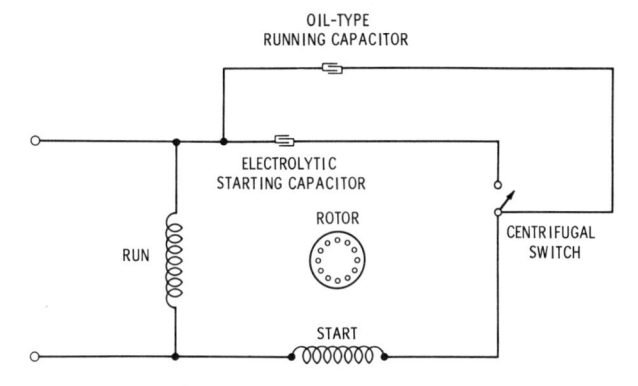

Fig. 14. A capacitor-start-capacitor-run motor.

12-61 Draw a schematic diagram showing the direction reversal of a capacitor-start motor. (See Fig. 15.)

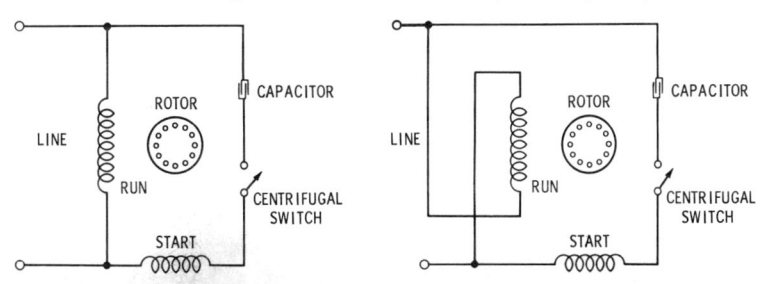

Fig. 15. The reversal of a capacitor-start motor is accomplished by reversing the leads of the running winding.

Motors

12-62 Draw a schematic diagram of a two-speed capacitor motor. (See Fig. 16.)

12-63 What is a shaded-pole motor?

This is a single-phase induction motor that is provided with one or more auxiliary short-circuited stator windings that are magnetically displaced from the main winding.

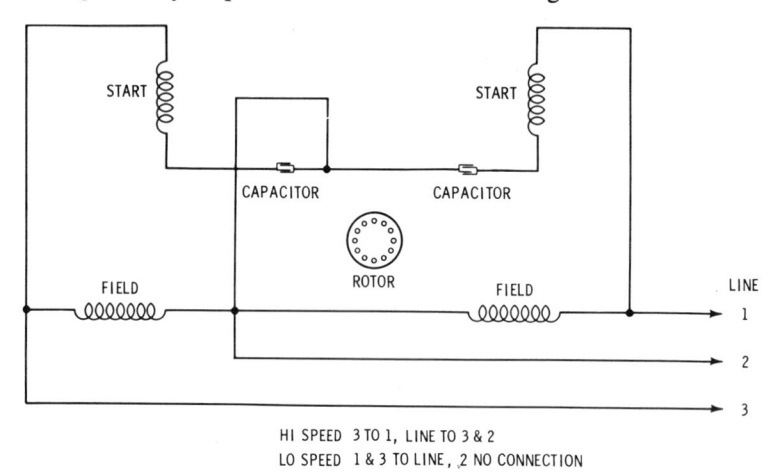

HI SPEED 3 TO 1, LINE TO 3 & 2
LO SPEED 1 & 3 TO LINE, 2 NO CONNECTION

Fig. 16. The schematic representation of a two-speed capacitor motor.

12-64 Does a shaded-pole motor have much starting torque?

No, it has very little starting torque, and for this reason it is used primarily for small fan motors.

12-65 In what direction do shaded-pole motors turn?

They rotate toward the shading coil.

12-66 Draw a sketch of one pole of a shaded-pole motor, showing the direction of rotation. (See Fig. 17.)

12-67 What is a repulsion motor?

This motor has a stator like that of most single-phase motors and a rotor similar to that of a DC motor. It has a field and an armature; however, the armature is not connected to the supply source, and the brushes are short circuited, or connected together, by a conductor of negligible resistance.

12-68 Draw a schematic diagram of a repulsion motor. (See Fig. 18.)

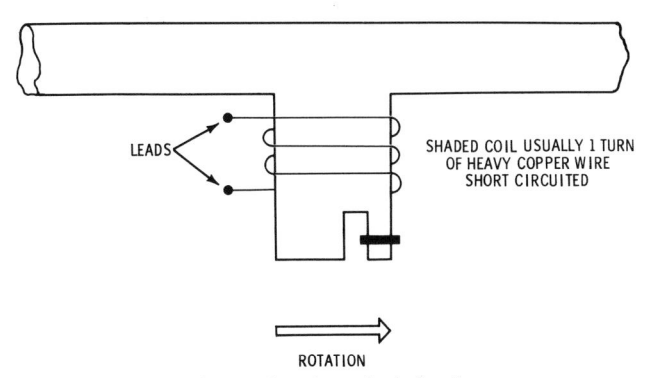

MOTORS

Fig. 17. One pole of a shaded-pole motor.

12-69 What is a repulsion-induction motor?
This is a repulsion motor with a squirrel-cage rotor on the armature, so that after the motor starts, it will run as an induction motor.

12-70 What is a repulsion-start-induction-run motor?
This is a repulsion motor that has a centrifugal-force device to short circuit the commutator when the motor reaches its rated speed.

12-71 How can the direction of rotation for the repulsion, repulsion-induction, and repulsion-start-induction-run motors be changed?

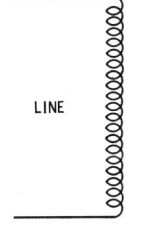

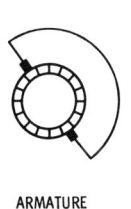

Fig. 18. The schematic representation of a repulsion motor.

By shifting the position of the brushes about 15° (electrical).

12-72 What is a universal motor?
This is a motor built like a series DC motor; however, the stator and armature are both laminated, they are designed for high speeds, and they may be used on either AC or DC, although the speed and power will be greater on DC.

229

MOTORS

12-73 How can the direction of rotation of a universal motor be changed?

By reversing either the field leads or the armature leads, but not both.

12-74 How can the direction of rotation of a DC motor be changed?

By reversing either the field or armature leads, but not both.

12-75 Knowing the frequency and the number of poles of an AC motor, how can its speed be determined? (This will be synchronous speed; the actual speed will be slightly lower.)

The speed of an AC motor can be found by using the formula

$$Speed = \frac{120 \times Frequency}{Number\ of\ poles}$$

12-76 We have a 60-cycle, 4-pole AC motor. What will its speed be?

$$Speed = \frac{120 \times 60}{4} = \frac{7{,}200}{4} = 1{,}800\ rpm$$

CHAPTER 13

Motor Controls

Motor controls are not really complicated. Once the fundamental idea is mastered, all types of controls can be figured out and sketched. First, you must become familiar with the common symbols that are used in connection with controls; second, analyze what you want to accomplish with the particular control. If you follow through, your diagram will fit right into place.

There are two common illustrative methods—diagrams and schematics. Both are used, and in practically every case, either one will be acceptable to the examiner. However, you should ask the examiner to be certain that you may use either; this gives you the opportunity to use the one you are most familiar with. The schematic method, however, presents a clearer picture of what you are trying to accomplish and is easier for the examiner to follow and correct. Whichever method you use, take your time and make a clear sketch; there is nothing so irritating to an examiner as trying to determine what your intent was and whether you had any idea of what you were doing.

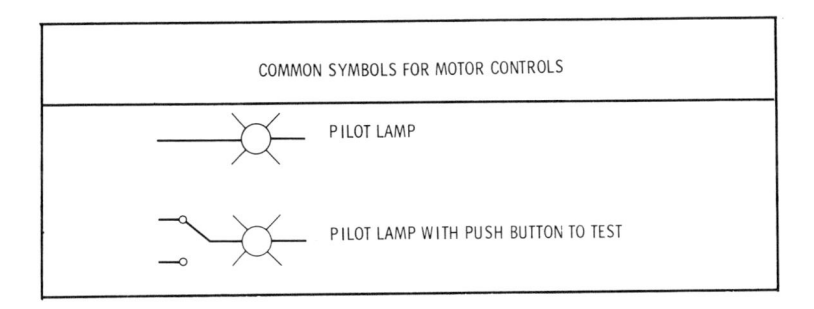

COMMON SYMBOLS FOR MOTOR CONTROLS
✳ PILOT LAMP
─o╱o─✳ PILOT LAMP WITH PUSH BUTTON TO TEST

Motor Controls

—⊣⊢z	NORMALLY OPEN CONTACTOR WITH BLOWOUT
—⊣/⊢z	NORMALLY CLOSED CONTACTOR WITH BLOWOUT
—⊣ ⊢—	NORMALLY OPEN CONTACTOR
—⊣/⊢—	NORMALLY CLOSED CONTACTOR
—⌒—	SHUNT COIL
—∧—	SERIES COIL
(coil + NC contact bracketed)	THERMAL OVERLOAD RELAY
⋛ ⧣	MAGNETIC RELAY
—◦╱◦—	LIMIT SWITCH, NORMALLY OPEN
—◦╲◦—	LIMIT SWITCH, NORMALLY CLOSED
—◦╱◦—	FOOT SWITCH, NORMALLY OPEN
—◦▬◦—	FOOT SWITCH, NORMALLY CLOSED
—◦T◦—	VACUUM SWITCH, NORMALLY OPEN
—◦⊤◦—	VACUUM SWITCH, NORMALLY CLOSED
—◦T◦—	LIQUID-LEVEL SWITCH, NORMALLY OPEN
—◦⊤◦—	LIQUID-LEVEL SWITCH, NORMALLY CLOSED

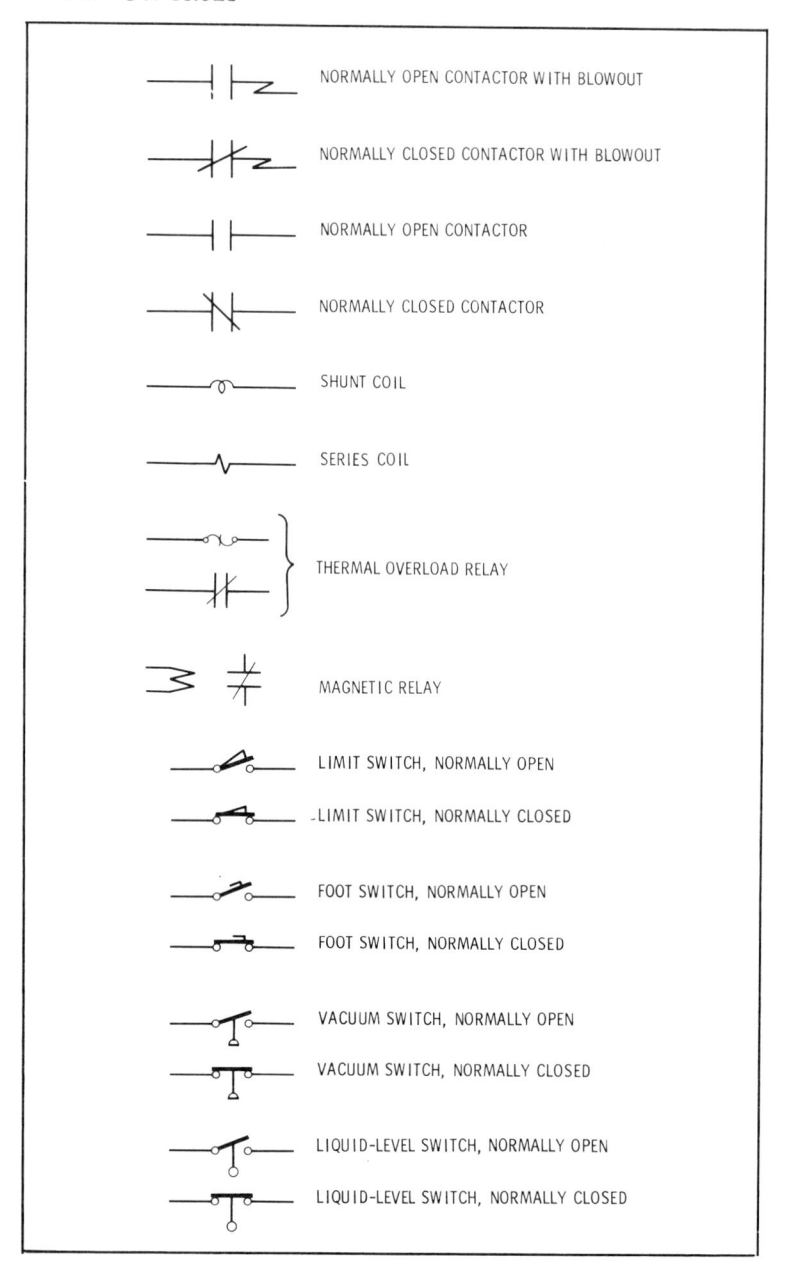

Motor Controls

TEMPERATURE ACTUATED SWITCH, NORMALLY OPEN

TEMPERATURE ACTUATED SWITCH, NORMALLY CLOSED

FLOW SWITCH, NORMALLY OPEN

FLOW SWITCH, NORMALLY CLOSED

MOMETARY-CONTACT SWITCH, NORMALLY OPEN

MOMETARY-CONTACT SWITCH, NORMALLY CLOSED

IRON-CORE INDUCTOR

AIR-CORE INDUCTOR

SINGLE-PHASE AC MOTOR

3-PHASE, SQUIRREL-CAGE MOTOR

2-PHASE, 4-WIRE MOTOR

WOUND-ROTOR, 3-PHASE MOTOR

ARMATURE

CROSSED WIRES, NOT CONNECTED

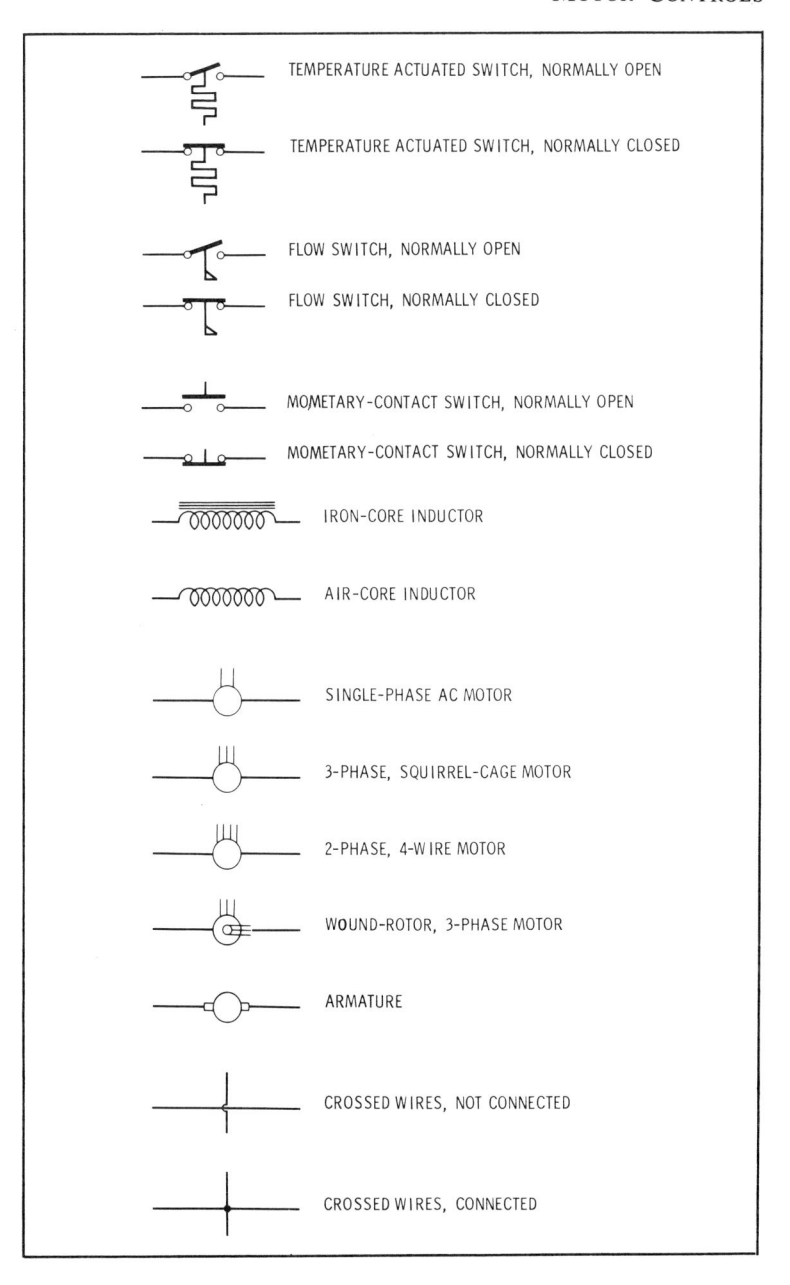

CROSSED WIRES, CONNECTED

MOTOR CONTROLS

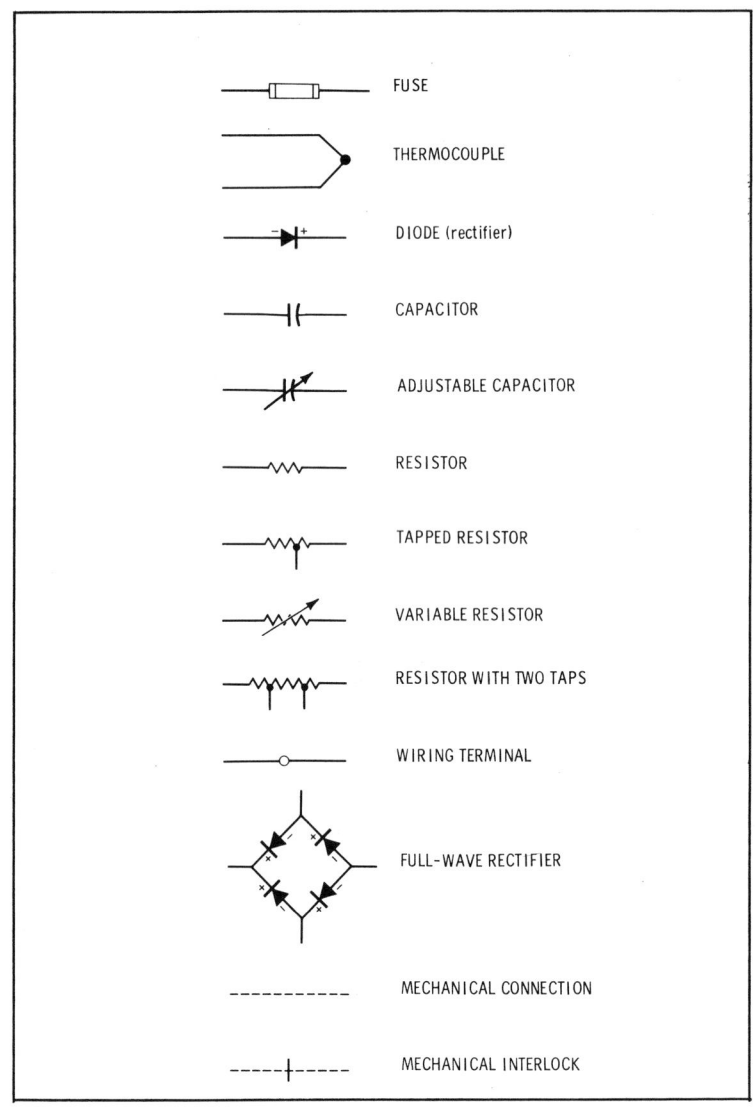

—▯—	FUSE
⊃•	THERMOCOUPLE
—▶︎|+—	DIODE (rectifier)
—||—	CAPACITOR
—⫲—	ADJUSTABLE CAPACITOR
—⏦—	RESISTOR
(tapped)	TAPPED RESISTOR
(variable)	VARIABLE RESISTOR
(two taps)	RESISTOR WITH TWO TAPS
—o—	WIRING TERMINAL
◇	FULL-WAVE RECTIFIER
- - - - - - -	MECHANICAL CONNECTION
- - -|- - -	MECHANICAL INTERLOCK

13-1 Draw a *diagram* of a magnetic three-phase starter with one start-stop station. (See Fig. 1.)

13-2 Draw a *schematic* of a magnetic starter with one start-stop station. (See Fig. 2.)

234

MOTOR CONTROLS

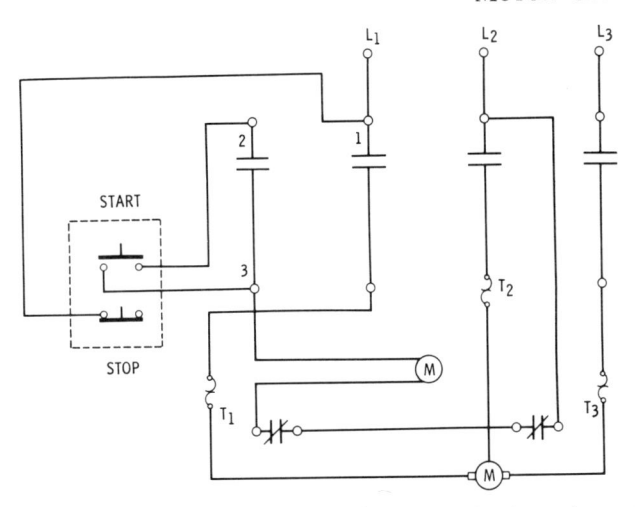

Fig. 1. A diagrammatic representation of a magnetic three-phase starter with one start-stop station.

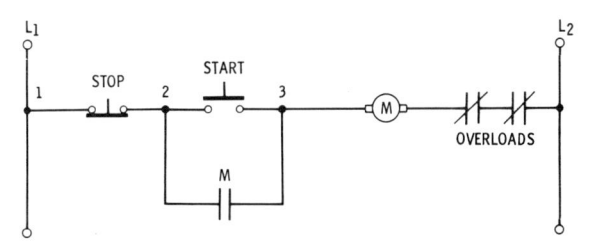

Fig. 2. A schematic representation of a magnetic three-phase starter with one start-stop station.

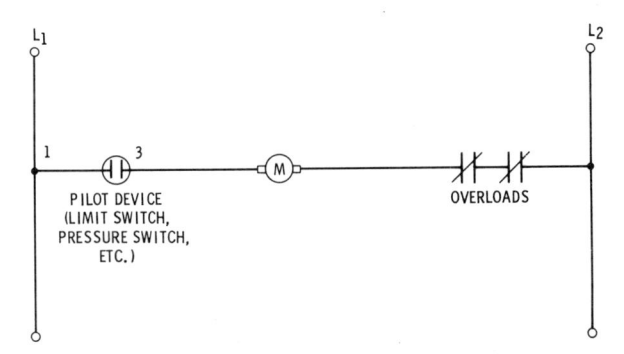

Fig. 3. A maintained-contact control.

235

Motor Controls

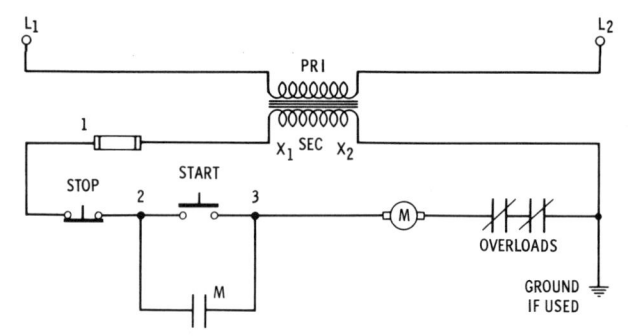

Fig. 4. A low-voltage motor control using a transformer.

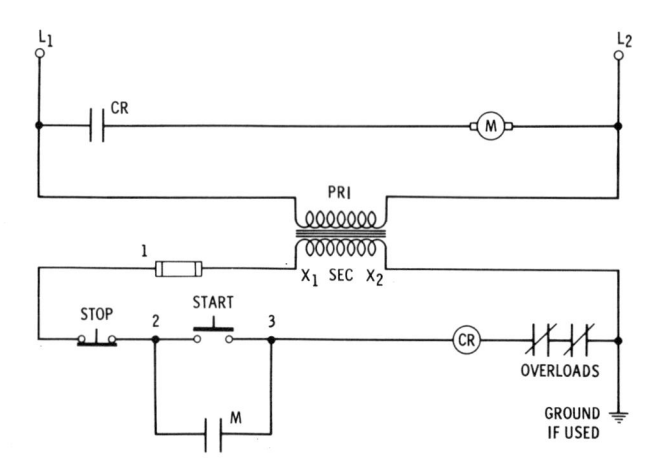

Fig. 5. A low-voltage motor control using a transformer in conjunction with a control relay.

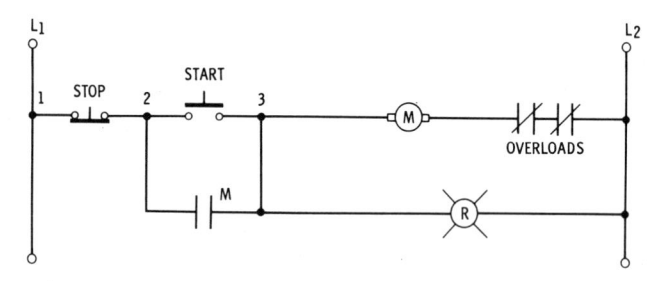

Fig. 6. A magnetic starter with one stop-start station and a pilot lamp which burns to indicate that the motor is running.

MOTOR CONTROLS

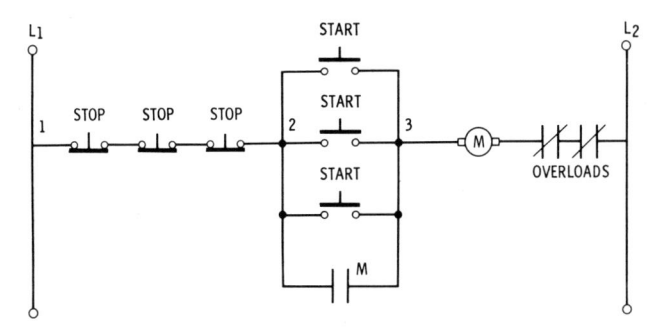

Fig. 7. A magnetic starter with three stop-start stations.

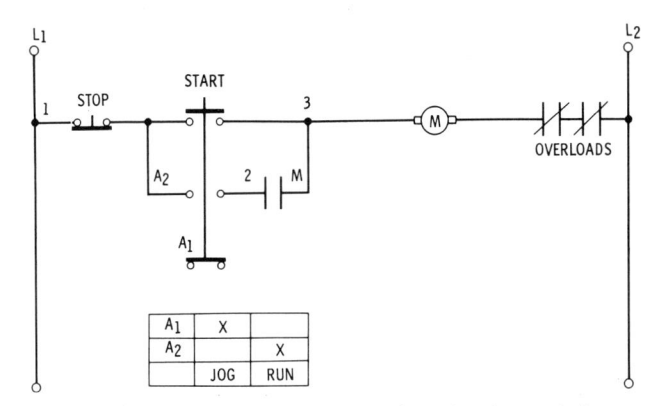

Fig. 8. A magnetic starter with a jogging switch.

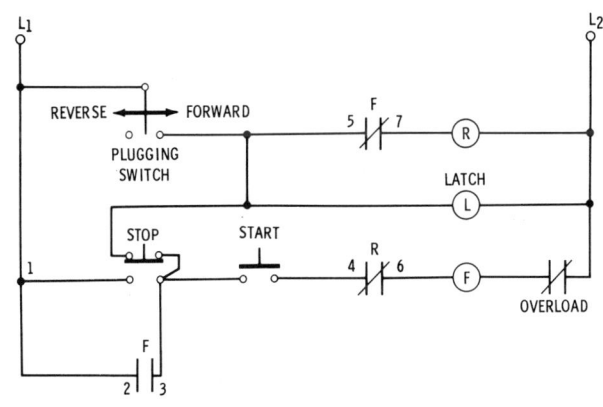

A_1	X	
A_2		X
	JOG	RUN

Fig. 9. A magnetic starter with a plugging switch and a safety latch.

237

Motor Controls

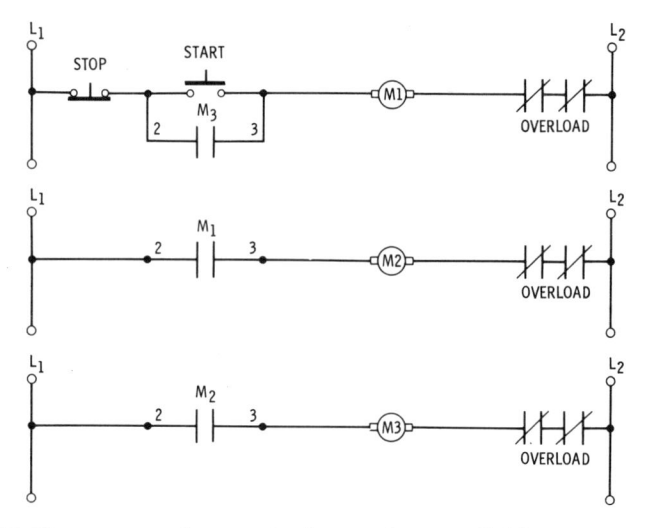

Fig. 10. Three motors that are simultaneously controlled by one stop-start station; if one overload device trips, all three motors will stop running.

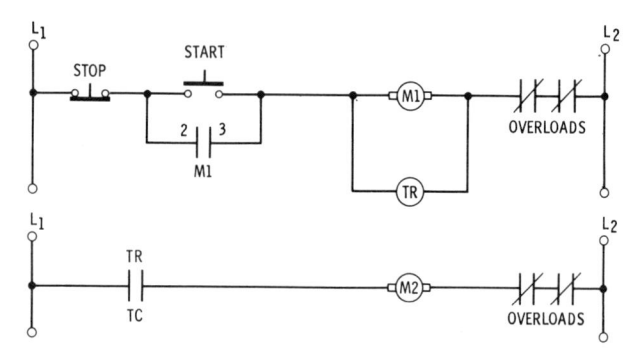

Fig. 11. Two magnetic starters that are controlled by one start-stop station; there is a starting-time delay device between the two motors.

Note that Figs. 1 and 2 both accomplish the same function, but you can see why the schematic method is preferred.

13-3 What happens on the normal magnetic starter if the electricity drops off for an instant?

The magnetic starter will drop out and will stay out until restarted.

MOTOR CONTROLS

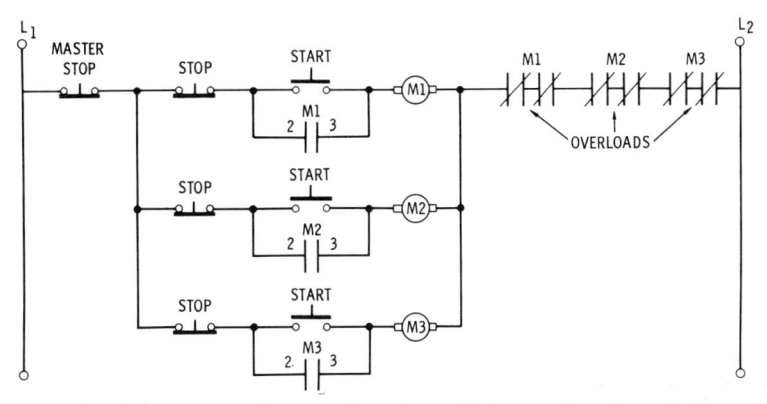

Fig. 12. Three separately started motors which may be stopped by one master stop station or stopped if one overload device is tripped.

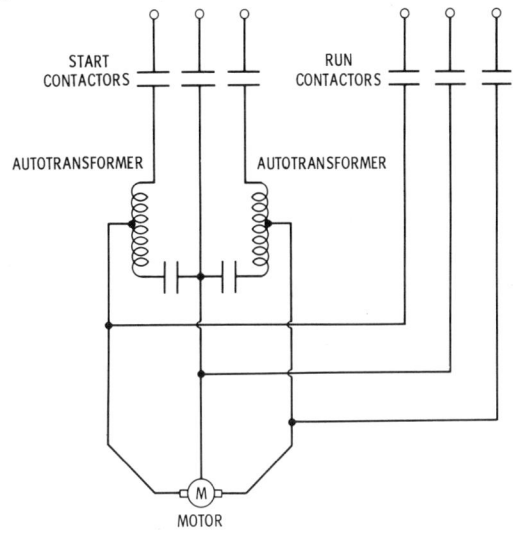

Fig. 13. A schematic representation of a starting compensator, with start and run contacts.

13-4 How can you restart the magnetic starter after a momentary lightning interruption?

The holding contacts (M in Fig. 2) are shunted with a timing device that maintains the contact for a certain time period, which is ordinarily adjustable for any desired delay.

13-5 In an area where a number of timing devices are used,

MOTOR CONTROLS

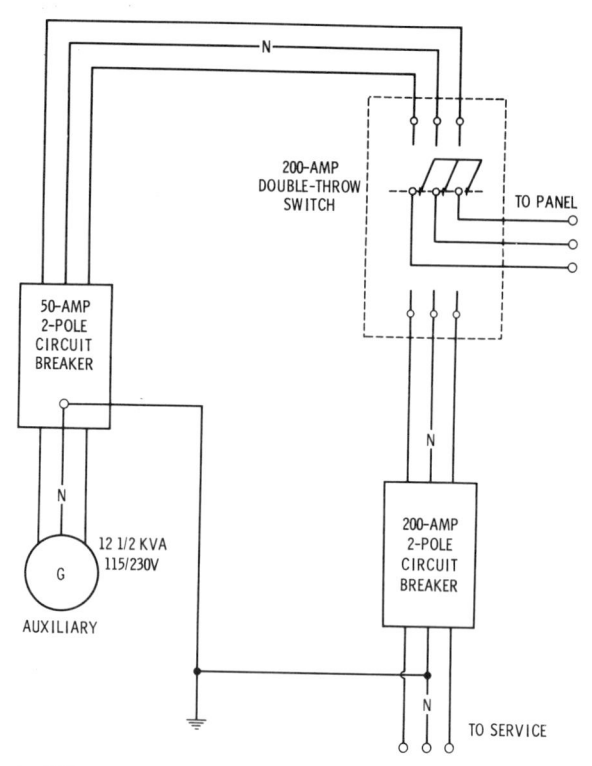

Fig. 14. A 200-ampere service with a 12½-KVA standby alternator arranged for a nonautomatic switch-over.

what would happen if they were all set for the same time-delay interval?

They would all come back on at the same instant, thereby causing a tremendous current surge on the distribution system, which would cause the line breakers to trip. This situation can be avoided by setting the time delays at different intervals of time, so that they will be staggered when coming back in the system.

13-6 Draw a maintained-contact control, and explain how it differs from Fig. 2.

The maintained-contact control has no voltage drop-out device; therefore, if the current is interrupted, the starter will come

240

MOTOR CONTROLS

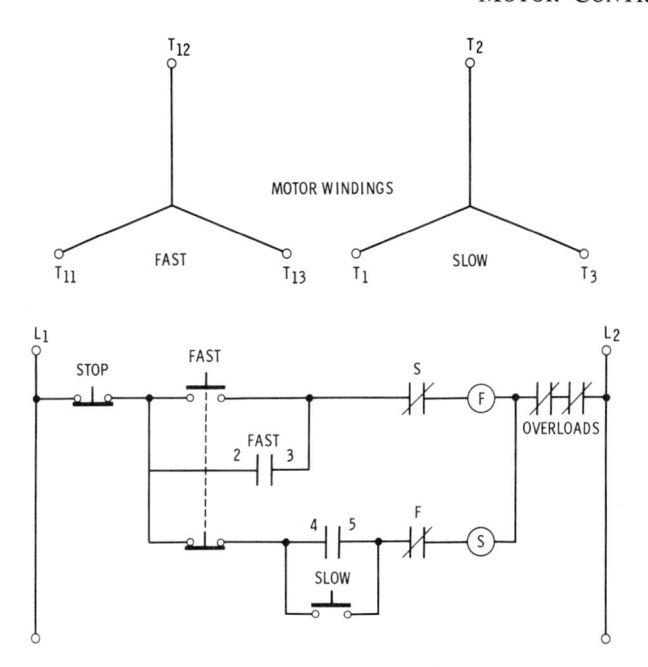

Fig. 15. A two-speed, three-phase, squirrel-cage motor starter.

back on the line when electric service is restored and will function normally until another interruption (see Fig. 3).

13-7 Name a few devices that may be used as the pilot in Fig. 3.

Limit switch, pressure switch, thermostat, flow switch, low-water switch, hand-operated switch.

13-8 How can you use low voltage to control a motor starter that is designed for use on high voltage?

A transformer or other source of voltage could be used, if the control wires are run in the same raceways as the motor conductors; they must have insulation equal to the proper voltage rating of the highest voltage used.

13-9 Draw a schematic of a low-voltage control using a transformer. (See Fig. 4.)

13-10 Draw a schematic of a low-voltage control using a control relay to energize the magnetic coil with full voltage. (See Fig. 5.)

MOTOR CONTROLS

13-11 Draw a schematic of a magnetic starter with one stop-start station and a pilot that burns when the motor is running. (See Fig. 6.)

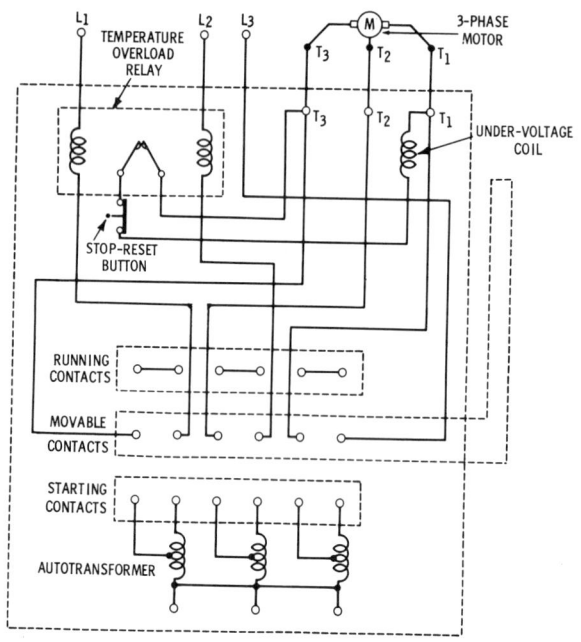

Fig. 16. A manual-type starting compensator.

13-12 Draw a schematic of a magnetic starter with three stop-start stations. (See Fig. 7.)

13-13 What is jogging?

This means inching a motor, or constant starting and stopping of the motor to move it a little at a time.

13-14 Draw a starter schematic with a jogging switch. (See Fig. 8.)

13-15 What is plugging?

This means stopping a motor by instantaneously reversing it until it stops.

13-16 Draw a starter schematic illustrating connections for plugging a motor provided with a safety latch. (See Fig. 9.)

13-17 Draw a schematic of three motors that are all

MOTOR CONTROLS

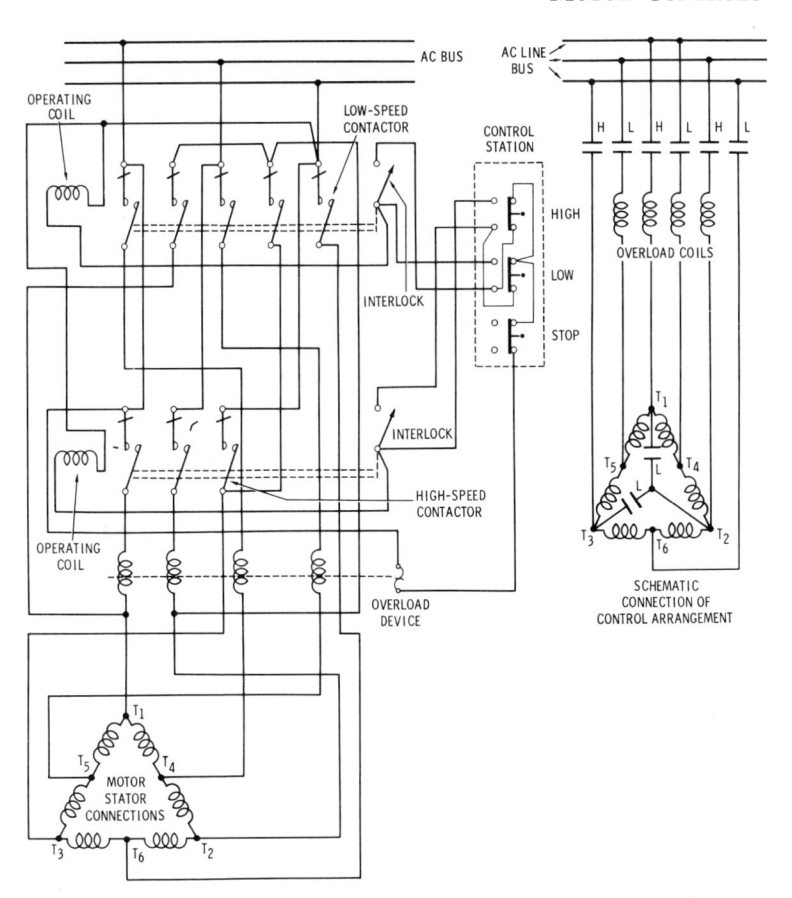

Fig. 17. A two-speed AC motor with push-button control.

started and stopped from one stop-start station so that they will all stop if one overload trips. (See Fig. 10.)

13-18 Draw a schematic of two magnetic starters controlled from one start-stop station, with a starting time delay between the two motors. (See Fig. 11.)

13-19 Draw a schematic of three separately started motors, all of which may be stopped by one master stop station or stopped if one overload trips. (See Fig. 12.)

13-20 Draw a schematic of a starting compensator, showing start and run contacts. (See Fig. 13.)

243

MOTOR CONTROLS

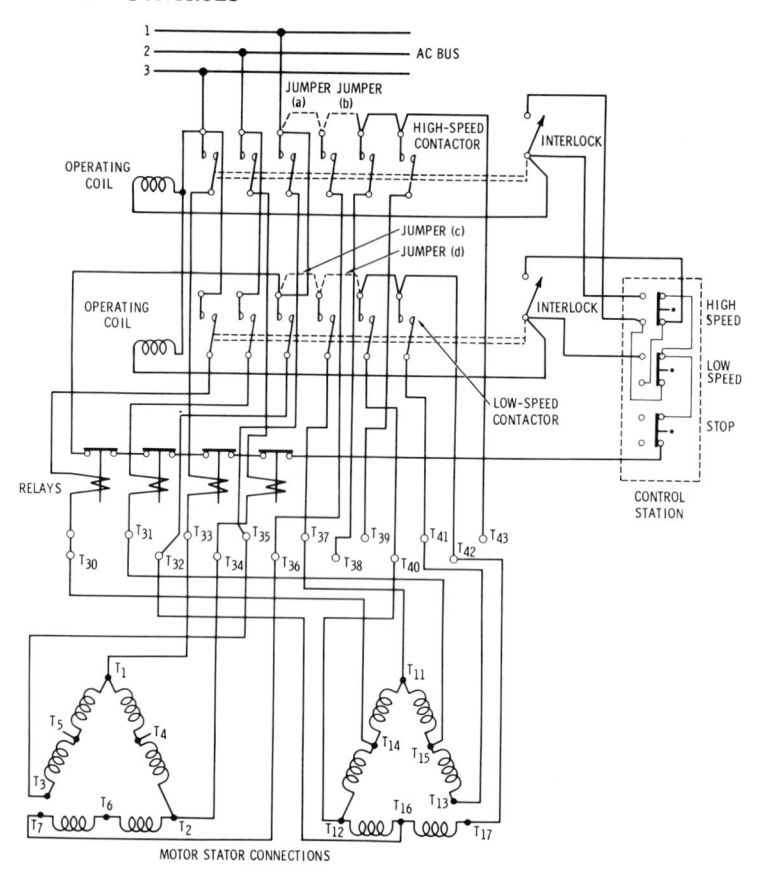

Fig. 18. The control arrangement for an AC multispeed motor.

13-21 Draw a diagram of a 200-ampere service with a 12½-KVA alternator as a standby; this will be a nonautomatic switch-over. The diagram must be arranged so that the two sources cannot be connected together, and both must have proper overload protection. (See Fig. 14.)

13-22 Draw a schematic of a two-speed, 3-phase, squirrel-cage motor starter; show motor connections. (See Fig. 15.)

13-23 What is a starting compensator?

MOTOR CONTROLS

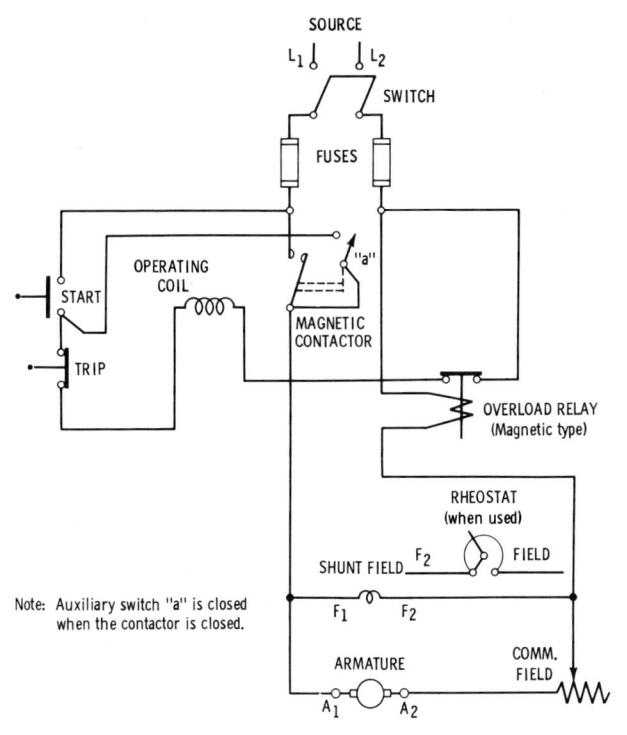

Fig. 19. The wiring diagram for a DC magnetic starter, with a contactor and an overload relay.

This is an AC device that consists of a built-in autotransformer, which reduces the voltage to the motor at start; after coming up to partial speed at a reduced voltage and reduced line current, the motor is connected across the line for its running position (see Fig. 13 for the magnetic-type compensator).

13-24 Draw a diagram of a manual-type starting compensator. (See Fig. 16.)

13-25 Draw a diagram of a two-speed AC motor control with push-button control. (See Fig. 17.)

13-26 Draw a diagram of the control arrangement for an AC multispeed motor. (See Fig. 18.)

13-27 Draw a wiring diagram for a DC magnetic starter with a contactor and overload relay. (See Fig. 19.)

Motor Controls

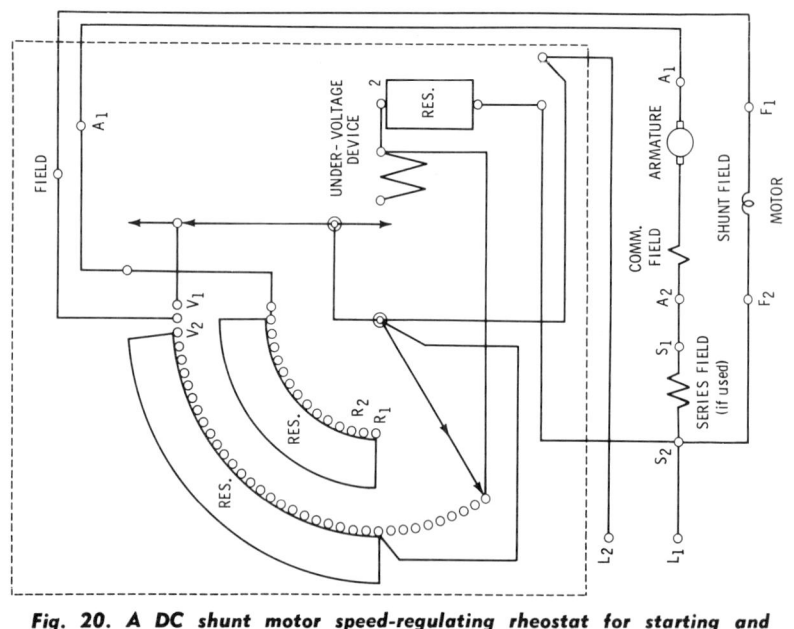

Fig. 20. A DC shunt motor speed-regulating rheostat for starting and speed control by field control.

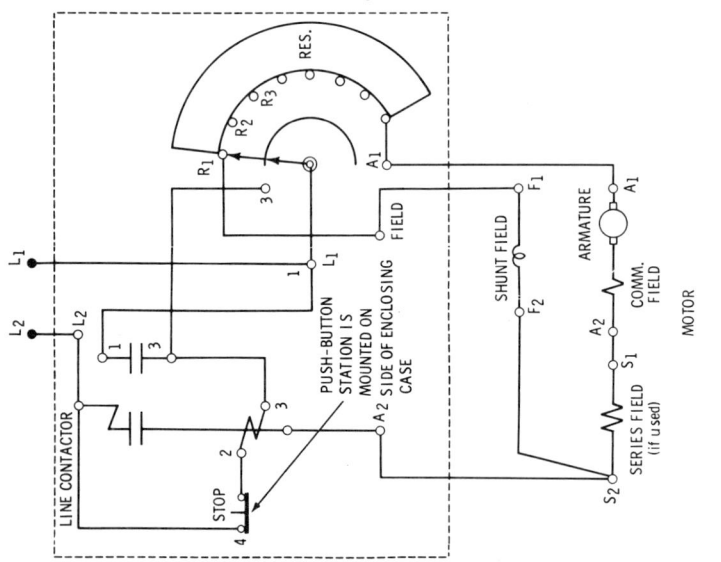

Fig. 21. A DC speed-regulating rheostat control for shunt or compound-wound motors, with contactors and push-button station control.

Motor Controls

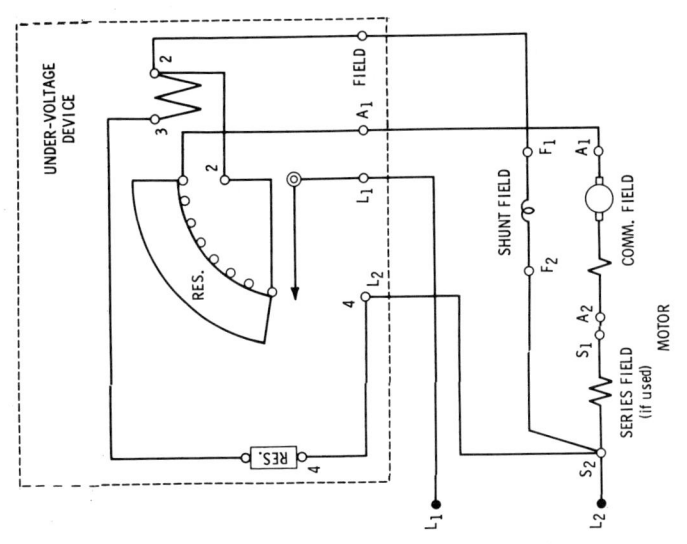

Fig. 22. A DC speed-regulating rheostat control for shunt or compound-wound motors, with a contactor.

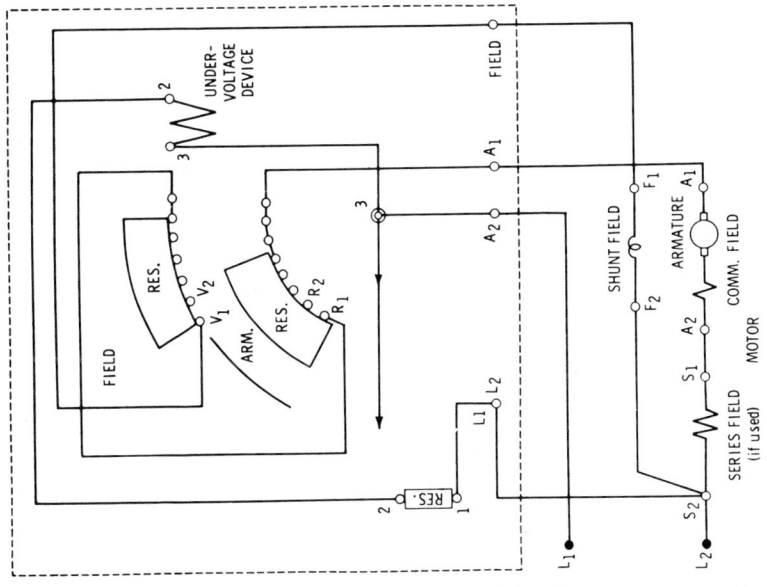

Fig. 23. A DC speed-regulating rheostat control for shunt or compound-wound motors; regulating duty—50% speed reduction by armature control and 25% increase by field control.

247

MOTOR CONTROLS

13-28 Draw a wiring diagram of a DC shunt motor speed-regulating rheostat for starting and speed control by field control. (See Fig. 20.)

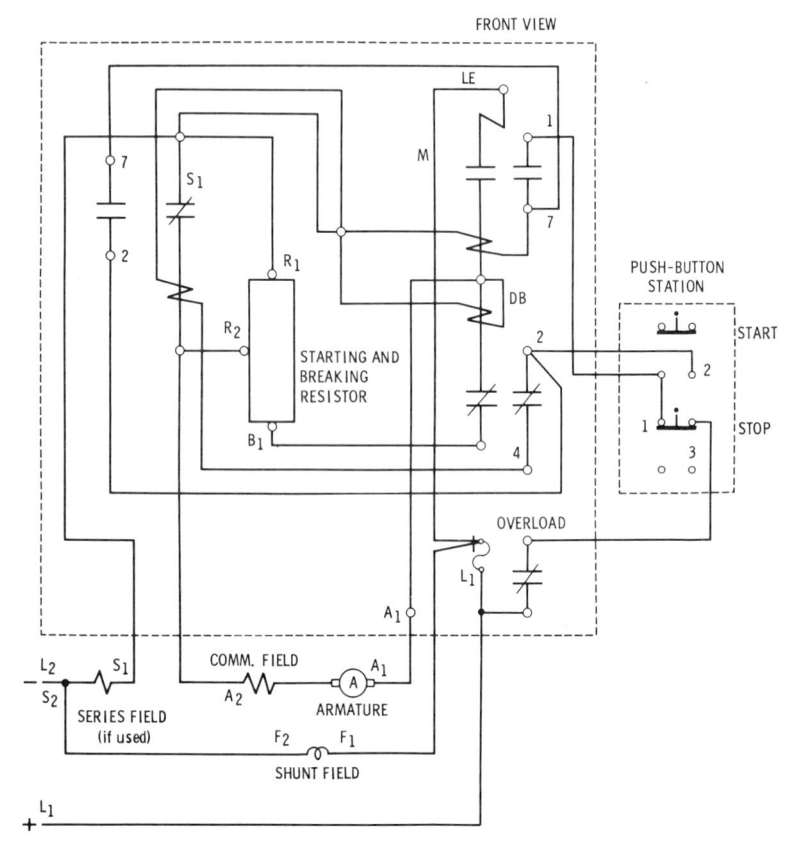

Fig. 24. A magnetic controller for constant-speed DC shunt or compound-wound nonreversible motors with dynamic breaking.

13-29 Draw a wiring diagram of a DC speed-regulating rheostat for shunt or compound-wound motors with contactors and push-button station control. (See Fig. 21.)

13-30 Draw a wiring diagram of a DC speed-regulating

248

MOTOR CONTROLS

rheostat for shunt or compound-wound motors without a contactor. (See Fig. 22.)

13-31 Draw a diagram of a DC speed-regulating rheostat for shunt or compound-wound motors. Regulating duty—50% speed reduction by armature control, and 25% increase by field control. (See Fig. 23.)

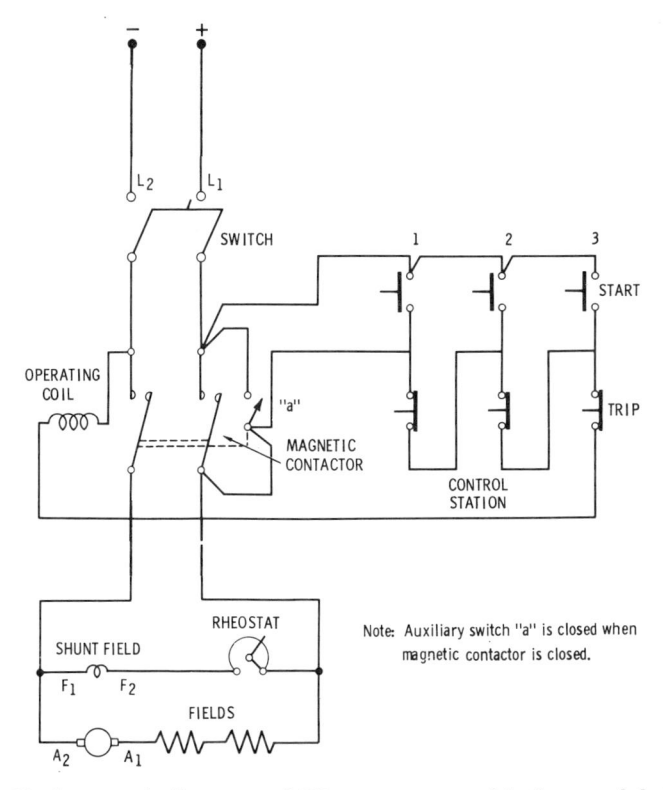

Fig. 25. A magnetically operated DC motor starter with three push-button control stations.

13-32 Draw a diagram of magnetic controller for a constant-speed DC shunt or compound-wound nonreversible motor with dynamic braking. (See Fig. 24.)

13-33 Draw a magnetically operated DC motor started with three push-button control stations. (See Fig. 25.)

249

MOTOR CONTROLS

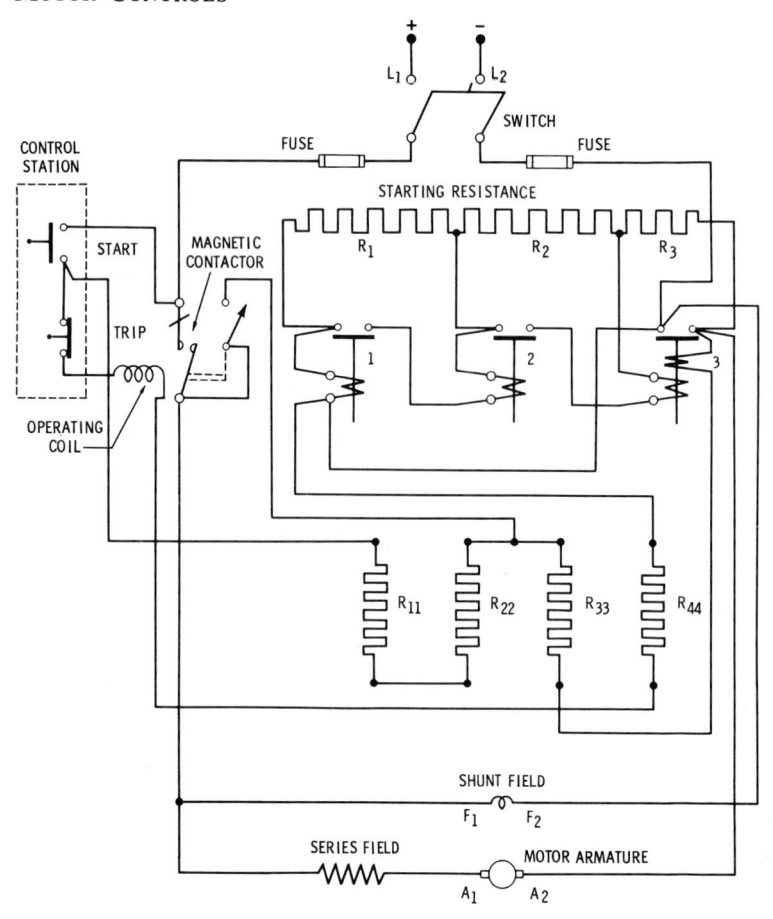

Fig. 26. *A push-button operated DC motor starter in which the armature starting current is limited by a step-by-step resistance regulation.*

13-34 Draw a push-button-operated DC motor starter, showing a starting arrangement wherein the armature starting current is limited by a step-by-step resistance regulation. (See Fig. 26.)

CHAPTER 14

Special Occupancies and Hazardous Locations

14-1 Where rigid conduit is used in hazardous locations, what precautions must be taken?

All threaded connections must be made up wrench tight to minimize sparking when fault currents flow through the conduit system. Where this is not possible, a bonding jumper must be used (see NEC, Article 500-1).

14-2 Various atmospheric mixtures require different treatment. What is the grouping given to these various mixtures?

The characteristics of various atmospheric mixtures of hazardous gases, vapors, and dusts depend on the specific hazardous material involved. It is necessary, therefore, that equipment be approved not only for the class of location but also for the specific gas, vapor, or dust that will be present.

For purposes of testing and approval, various atmospheric mixtures have been grouped on the basis of their hazardous characteristics, and facilities have been made available for testing and approval of equipment for use in the following atmospheric groups:

Group A—Atmospheres containing acetylene;

Group B—Atmospheres containing hydrogen, or gases or vapors of equivalent hazard such as manufactured gas;

Group C—Atmospheres containing ethylether vapors, ethylene, or cyclopropane;

Group D—Atmospheres containing gasoline, hexane,

naphtha, benzine, butane, propane, alcohol, acetone, benzol, lacquer solvent vapors, or natural gas;

Group E—Atmospheres containing metal dust, including aluminum, magnesium, and their commercial alloys, and other metals of similarly hazardous characteristics;

Group F—Atmospheres containing carbon black, coal, or coke dust;

Group G—Atmospheres containing flour, starch, or grain dusts.

14-2a How shall equipment for hazardous locations be marked?

Approved equipment shall be marked to show the class, group and operating temperature, or temperature range, based on operation in a 40°C ambient for which it is approved.

The temperature range, if provided, shall be indicated in identification numbers as shown in Table 1.

Identification numbers marked on the equipment nameplates shall be in accordance with Table 1.

Exception: Equipment of the nonheat-producing type, such as junction boxes, conduit and fittings, are not required to have a marked operating temperature.

For purpose of testing and approval, various atmospheric mixtures (not oxygen enriched) have been grouped on the basis of their hazardous characteristics, and facilities have been made available for testing and approval of equipment for use in the atmospheric groups listed in Table 2. Since there is no consistent relationship between explosion properties and ignition temperature, the two must be regarded as independent requirements.

Table 1. Identification Numbers

Maximum Temperature Degrees C	Degrees F	Identification Number
450	842	T1
300	572	T2
280	536	T2A
260	500	T2B
230	446	T2C

Special Occupancies and Hazardous Locations

Table 1. Identification Numbers

Maximum Temperature Degrees C	Degrees F	Identification Number
215	419	T2D
200	392	T3
180	356	T3A
165	329	T3B
160	320	T3C
135	275	T4
120	248	T4A
100	212	T5
85	185	T6

Table 2. Chemicals by Groups

Group A Atmospheres

Chemical

acetylene

Group B Atmospheres

Butadiene[1]
ethylene oxide[2]
hydrogen
manufactured gases containing more than 30% hydrogen (by volume)
propylene oxide[2]

Group C Atmospheres

acetaldehyde
cyclopropane
diethyl ether
ethylene
isoprene
unsymmetrical dimethyl hydrazine (UDMH 1, 1-dimethyl hydrazine)

Group D Atmospheres

Chemical

acetone
acrylonitrile
ammonia[3]
benzene
butane
1-butanol (butyl alcohol)
2-butanol (secondary butyl alcohol)
n-butyl acetate
isobutyl acetate
ethane
thanol (ethyl alcohol)
ethyl acetate
ethylene dichloride
gasoline
heptanes
hexanes
isoprene
methane (natural gas)
methanol (methyl alcohol)
3-methyl-1-butanol (isoamyl alcohol)
methyl ethyl ketone
methyl isobutyl ketone
2-methyl-1-propanol (isobutyl alcohol)
2-methyl-2-propanol (tertiary butyl alcohol)
petroleum naphtha[4]

Special Occupancies and Hazardous Locations
Table 2. Chemicals by Groups

Group A Atmospheres Chemical	Group D Atmospheres Chemical
	octanes
	pentanes
	1-pentanol (amyl alcohol)
	propane
	1-propanol (propyl alcohol)
	2-propanol (isopropyl alcohol)
	propylene
	styrene

[1] Group D equipment may be used for this atmosphere if such equipment is isolated in accordance with Section 501-5(a) by sealing all conduit ½-inch size or larger.

[2] Group C equipment may be used for this atmosphere if such equipment is isolated in accordance with Section 501-5(a) by sealing all conduit ½-inch size or larger.

[3] For Classification of areas involving ammonia atmosphere refer to ANSI B9.1 Safety Code for Mechanical Refrigeration-1971 and ANSI K61.1 Storage and Handling of Anhydrous Ammonia—1975.

[4] A saturated hydrocarbon mixture boiling in the range 20-135°C (68-275°F). Also known by the synonyms benzine, ligroin, petroleum ether or naphtha.

14-3 What is a Class I location?

These are locations in which flammable gases or vapors are, or may be, present in the air in quantities sufficient to produce explosive or ignitible mixtures (see NEC, Article 500-4).

14-4 What is a Class I, Division 1 location?

Locations in which hazardous concentrations of flammable gases or vapors exist continuously, intermittently, or periodically under normal operating conditions; in which hazardous concentrations of such gases or vapors may exist frequently because of repair or maintenance operations or because of leakage; or in which breakdown or faulty operations of equipment or processes that might release hazardous concentrations of flammable gases or vapors might also cause simultaneous failure of electrical equipment (see NEC, Article 500-4).

14-5 What is a Class I, Division 2 location?

Locations in which flammable volatile liquids or flammable gases are handled, processed, or used, but in which the hazardous liquids, vapors, or gases will normally be confined within closed containers or closed systems from which they can escape

Special Occupancies and Hazardous Locations

only in case of accidental rupture or breakdown of such containers or systems, or in case of abnormal operation of equipment; in which hazardous concentrations of gases or vapors are normally prevented by positive mechanical ventilation, but which might become hazardous through failure or abnormal operation of the ventilating equipment; or which are adjacent to Class I, Division 1 locations and to which hazardous concentrations of gases or vapors might occasionally be communicated unless such communication is prevented by adequate positive-pressure ventilation from a source of clean air, and effective safeguards against ventilation failure are provided (see NEC, Article 500-4).

14-6 What are Class II locations?

Those which are hazardous because of the presence of combustible dust (see NEC, Article 500-5).

14-7 What are Class II, Division 1 locations?

Locations in which combustible dust is, or may be, in suspension in the air continuously, intermittently or periodically under normal operating conditions, in quantities sufficient to produce explosive or ignitible mixtures; where mechanical failure or abnormal operation of machinery or equipment might cause such mixtures to be produced, and might also provide a source of ignition through simultaneous failure of electrical equipment, operation of protective devices, or from other causes; or in which dusts of an electrical conducting nature may be present (see NEC, Article 500-5).

14-8 What are Class II, Division 2 locations?

Locations in which combustible dust will not normally be in suspension in the air, or will not be likely to be thrown into suspension by the normal operation of equipment or apparatus, in quantities sufficient to produce explosive or ignitible mixtures, but where deposits or accumulations of such dust may be sufficient to interfere with the safe dissipation of heat from electrical equipment and apparatus, or where such deposits or accumulations of dust on, in, or in the vicinity of electrical equipment might be ignited by arcs, sparks, or burning material from such equipment (see NEC, Article 500-5).

14-9 What are Class III locations?

Locations which are hazardous because of the presence of easily ignitible fibers or flyings, but in which such fibers or flyings

Special Occupancies and Hazardous Locations

are not likely to be suspended in the air in quantities sufficient to produce ignitible mixtures (see NEC, Article 500-6).

14-10 What are Class III, Division 1 locations?

Locations in which easily ignitible fibers or materials producing combustible flyings are handled, manufactured, or used (see NEC, Article 500-6).

14-11 What are Class III, Division 2 locations?

Locations in which easily ignitible fibers are stored or handled, except in process or manufacture of the product (see NEC, Article 500-6).

14-12 Where may meters, instruments, and relays be mounted in Class I, Division 1 locations?

Within enclosures approved for this location (see NEC, Article 501-3).

14-13 Where may meters, instruments, and relays with make-and-break contacts be mounted in Class I, Division 2 locations?

Immersed in oil or hermetically sealed against vapor or gases (see NEC, Article 501-3).

14-14 What type of wiring must be used in Class I, Division 1 areas?

Rigid metal conduit (threaded) or Type MI cable (see NEC, Article 501-4).

14-15 What type of fittings must be used in Class I, Division 1 areas?

All boxes, fittings, and joints must be threaded. At least five full threads must be used and must be fully engaged, and all materials, including flexible connections, must be approved (explosion-proof) by the inspecting authority for these locations (see NEC, Article 501-4).

14-16 What type of wiring must be used in Class I, Division 2 areas?

Rigid metal conduit (threaded) or Types MI, MC, TC, or ALS SNM Cable with approved termination fittings. (See NEC, Section 501-4.)

14-17 What type of fittings must be used in Class 1, Division 2 areas?

Only boxes, fittings, and joints that are approved for this location (see NEC, Article 501-4).

Special Occupancies and Hazardous Locations

14-18 Why is sealing needed in conduit systems and not in Type MI cable systems in Class I areas?

Seals are provided in conduit systems to prevent the passage of gases, vapors, or fumes from one portion of the electrical system to another through the conduit. Type MI cable is inherently constructed to prevent the passage of gases, etc., but sealing compound is used in cable-termination fittings to exclude moisture and other fluids from the cable insulation (see NEC, Article 501-5).

14-19 When connecting conduit to switches, circuit breakers, etc., where must seals be placed?

As close as possible to the enclosure, but not more than 18 inches away (see NEC, Article 501-5).

14-20 When a conduit leaves a Class I, Division 1 location, where must the seal be located?

The seal must be at the first fitting when a conduit leaves this area (see NEC, Article 501-5). On either side of the boundary.

14-21 When conduit leaves a Class I, Division 2 location, where must the seal be located?

The seal must be at the first fitting when a conduit leaves this area; it may be on either side of the boundary (see NEC, Article 501-5).

14-22 If there is a chance that liquid will accumulate at a seal, what precautions must be taken?

An approved seal for periodic draining of any accumulation must be provided (see NEC, Article 501-5).

14-23 What must be provided for switches, motor controllers, relays, fuses, or circuit breakers in Class I locations?

They must be provided with enclosures and must be approved for the location in which they are located. UL listings are recommended for this purpose, and most enforcing agencies use this listing (see NEC, Article 501-6).

14-24 What is the required thickness of the sealing compound in Class I locations?

Not less than the trade size of the conduit, and in no case less than ⅝-inch thickness (see NEC, Article 501-5).

14-25 Must lighting fixtures for Class I locations be approved fixtures?

257

Special Occupancies and Hazardous Locations

They definitely must be approved for the classification in which they are used, and must be so marked; portable lamps must also be approved for these areas (see NEC, Article 501-9).

14-26 Is grounding necessary in Class I locations?

Yes, it is highly important. All exposed noncurrent-carrying parts of equipment are required to be grounded. Locknuts and bushings are not adequate grounding; they must have bonding jumpers around them. Where flexible conduit is used as permitted, bonding jumpers must be provided around such conduit (see NEC, Article 501-16).

14-27 Is lightning protection required in Class I locations?

Yes, lightning-protection devices are required on all ungrounded conductors; they must be connected ahead of the service-disconnecting means and must be bonded to the raceway at the service entrance (see NEC, Article 501-16).

14-28 What are some of the precautions that must be taken when transformers are used in Class II, Division 1 locations?

Transformers containing a flammable liquid must be installed only in approved vaults which are so constructed that a fire cannot be communicated to the hazardous area. Transformers that do not contain a flammable liquid must also be installed in vaults or must be enclosed in tight metal housings without ventilation (see NEC, Article 502-2).

14-29 What type of wiring must be installed in Class II, Division 1 locations?

Rigid metal conduit (threaded) or Type MI cable must be used. Boxes and fittings must have threaded bosses and tight-fitting covers, and must be approved for the location (see NEC, Article 502-4).

14-30 What type of wiring must be installed in Class II, Division 2 locations?

Rigid metal conduit, electrical metallic tubing, or Type MI, MC, ALS, or SNM Cable. Boxes, fitting, and joints must be made to minimize the entrance of dust (see NEC, Article 502-4).

14-31 What is needed for sealing in Class II locations?

Where a raceway provides communication between an enclosure that is required to be dust-ignition-proof and one that is

Special Occupancies and Hazardous Locations

not, a suitable means must be provided to prevent the entrance of dust into the dust-ignition-proof enclosure through the raceway. This means may be a permanent and effective seal, a horizontal section of raceway not less than 10 feet long, or a vertical section of raceway not less than 5 feet long and extending downward from the dust-ignition-proof enclosure (see NEC, Article 502-5).

14-32 What type of fixtures must be provided for Class II locations?

The fixtures must be approved for these locations (see NEC, Section 502-11).

14-33 Is there a difference between Class I fixtures and Class II fixtures?

Class I fixtures must be vapor-proof and capable of withstanding and containing an explosion from within. Class I vapors usually have a higher flash point than dusts in a Class II location; therefore, the glass enclosure on Class I fixtures, while it must be heavier to withstand an explosion from within, may be smaller because of the higher flash temperatures. Class II fixtures are faced with a heat-dissipation problem, because grain dusts and other types of dusts have a low flash-point temperature; therefore, the glass enclosure on Class II fixtures must not be allowed to reach a high temperature.

14-34 Is grounding and bonding necessary in Class II locations?

Yes. All exposed noncurrent-carrying parts of equipment are required to be grounded. Locknuts and bushings are not adequate grounding; they must use bonding jumpers (see NEC, Article 502-16).

14-35 Is lightning protection required in Class II locations?

Yes, lightning-protection devices of the proper type are required on all ungrounded conductors; they must be connected ahead of the service-disconnecting means and must be bonded to the raceway at the service entrance (see NEC, Article 502-16).

14-36 What type of wiring is required in Class III, Division 1 locations?

Rigid metal conduit or Types MI, MC, or ALS Cable; boxes

Special Occupancies and Hazardous Locations

and fittings must have tight-fitting covers. There must not be any screw-mounting holes within the box through which sparks might escape (see NEC, Article 503-3).

14-37 What vehicles are included under the commercial-garages classification? How many vehicles will put a garage in this classification?

These locations include places of storage, repairing, or servicing of self-propelled vehicles, including passenger automobiles, busses, trucks, tractors, etc., in which flammable liquids or flammable gases are used for fuel or power. (see NEC, Article 511-1).

14-38 What are the various area classifications in commercial garages?

For each floor at or above grade, the entire area is considered as a Class I, Division 2 location to a level of 18 inches above the floor. When below grade, the entire area to 18 inches above grade is considered as a Class I, Division 2 location, unless the area has positive ventilation, in which case it will be 18 inches above each such floor. Any pit or depression in the floor may be considered a Class I, Division 1 location by the enforcing authority. Adjacent stock rooms, etc. must be considered as Class I, Division 2 locations, unless there is a tight curb or elevation of 18 inches above the hazardous area (see NEC, Article 511-2).

14-38a In commercial garages, how should areas adjacent to the hazardous area be classified?

Adjacent areas which by reason of ventilation, air pressure differentials or physical spacing are such that in the opinion of the authority enforcing this Code no hazard exists, shall be classified as non-hazardous.

14-39 What material may be used in wiring commercial garages?

Rigid metal conduit or Type MI, MC, ALS, or SNM cable.

14-40 What is required above the hazardous areas for all fixed wiring in commercial garages?

Metallic raceways, Type MI cable, or Type ALS cable (see NEC, Article 511-5).

14-41 Where are seals required in commercial garages?

Special Occupancies and Hazardous Locations

Wherever conduit passes from Class I, Division 2 areas to nonhazardous areas; this applies to horizontal as well as vertical boundaries of these areas (see NEC, Article 511-4).

14-42 What is required on equipment above the hazardous area in commercial garages?

Equipment that is less than 12 feet above the floor, and which might produce sparks or particles of hot metal, must be totally enclosed to prevent the escape of sparks or hot metal particles, or be provided with guards or screens for the same purpose. Lighting fixtures must be located not less than 12 feet above the floor level unless protected by guards, screens, or covers (see NEC, Article 511-6).

14-47 What classification do pits or depressions of aircraft hangars require?

These are considered as Class I, Division 1 locations (see NEC, Article 513-2).

14-48 What is the normal classification of an aircraft hangar?

The entire area of a hangar, including adjacent areas not suitably cut off from the hangar, is considered as a Class I, Division 2 location to a height of 18 inches above the floor. Areas within 5 feet, horizontally, from aircraft power plants, aircraft fuel tanks, or aircraft structures containing fuel are considered as Class I, Division 2 locations to a point extending upward from the floor to a level 5 feet above the upper surfaces of wings and engine enclosures (see NEC, Article 513-2).

14-49 What type of wiring is required in all aircraft-hangar locations?

The wiring must be in accordance with Class I, Division 1 locations (see NEC, Article 513-3).

14-50 What type of wiring may be used in areas that are not considered as hazardous in aircraft hangars?

Metallic raceway, Type MI cable, TC, SNM, MC or Type ALS cable (see NEC, Article 513-43.

14-51 How must equipment, including lighting, above the aircraft be treated?

Any equipment or lighting less than 10 feet above wings and engine enclosures must be enclosed or suitably guarded to pre-

Special Occupancies and Hazardous Locations

vent the escape of arcs, sparks, or hot metal particles (see NEC, Article 513-5).

14-52 Is sealing required in aircraft hangars?

Yes, where wiring extends into or from hazardous areas; this regulation includes horizontal as well as vertical boundaries. Any raceways in the floor or below the floor are considered as being in the hazardous areas (see NEC, Article 513-7).

14-53 Is grounding required in aircraft hangars?

Yes, all raceways and all noncurrent-carrying metallic portions of fixed or portable equipment, regardless of voltage, must be grounded (see NEC, Article 513-12).

14-54 What is a gasoline dispensing and service station?

This includes locations where gasoline or other volatile flammable gases are transferred to the fuel tanks (including auxiliary fuel tanks) of self-propelled vehicles. Lubritoriums, service rooms, repair rooms, offices, salesrooms, compressor rooms, and similar locations must conform to Article 511 of the NEC (see NEC, Article 514-1).

14-55 What are the hazardous areas in and around gasoline dispensing islands?

The area within the dispenser and extending for a distance of 18 inches in all directions from the enclosure and extending upward for a height of 4 feet from the driveway; any wiring within or below this area will be considered as a Class I, Division 1 location and must be approved for this location. Any area in an outside location within 20 feet, horizontally, from the exterior enclosure of any dispensing pump is considered as a Class I, Division 2 location. Any areas within buildings not suitably cut off from this 20-foot area must also be considered as a Class I, Division 2 location. The Class I, Division 2 location within the 20-foot area around dispensing pumps extends to a depth of 18 inches below the driveway or ground level. Any area in an outside location within 10 feet, horizontally, from any tank fill-pipe must be considered as a Class I, Division 2 location, and any area within a building in the 10-foot radius must be considered the same. Electrical wiring and equipment emerging from dispensing pumps must be considered as a Class I, Division 1 location, at least to its point of emergence from this area. Any area within a 3-foot radius of the point of discharge of any tank vent pipe

Special Occupancies and Hazardous Locations

must be considered as a Class I, Division 1 location, and below this point, the area must be considered as a Class I, Division 2 location (see NEC, Article 514-2). A new part has been added in the 1971 NEC to Article 514-2, which is as follows:

Where the dispensing unit, including the hose and hose nozzle valve, is suspended from a canopy, ceiling or structural support, the Class I, Division 1 location shall include the volume within the enclosure and shall extend 18 inches in all directions from the enclosure where not suitably cutoff by a ceiling or wall. The Class I, Division 2 location shall extend 2 feet horizontally in all directions beyond the Division 1 classified area and extend to grade below this classified area. In addition, the horizontal area 18 inches above grade for a distance of 20 feet, measured from a point vertically below the edge of any dispenser enclosure, shall be classified Division 2. All electrical equipment integral with the dispensing hose and nozzle shall be suitable for use in a Division 1 location. See NEC, Section 514-2.

14-56 What are the restrictions on switching circuits leading to or going through a gasoline-dispensing pump?

The switches or circuit breakers in these circuits must be able to simultaneously disconnect all conductors of the circuit, including the grounded neutral, if any (see NEC, Article 514-5).

14-57 Are seals required in service-station locations?

Yes, approved seals must be provided in each conduit run entering or leaving a gasoline-dispensing pump or other enclosure located in a Class I, Division 1 or Division 2 location when connecting conduit originates in a nonhazardous location. The first fitting after the conduit emerges from the slab or from the concrete must be a sealing fitting (see NEC, Article 514-6).

14-58 Is grounding required in a service station wiring system?

Yes, all metal portions of dispensing islands, all metallic raceways, and all noncurrent-carrying parts of electrical equipment must be grounded, regardless of voltage (see NEC, Article 514-7).

14-59 What are bulk-storage plants?

Locations where gasoline or other volatile flammable liquids are stored in tanks having an aggregate capacity of one carload or more, and from which such products are distributed, usually by tank truck (see NEC, Article 515-1).

14-60 What are the classifications of bulk-storage plants?

Adequately ventilated indoor areas containing pumps, bleeders, withdrawal fittings, meters, and similar devices are considered as Class I, Division 2 locations in an area extending 5 feet in all directions from the exterior surfaces of such equipment; this location also extends 25 feet horizontally from any surface of the equipment and 3 feet above the floor or grade level. Areas that are not properly ventilated are required to have the same distances as above, but are considered as Class I, Division 1 locations (see NEC, Article 515-2).

14-61 Is underground wiring permitted in bulk-storage plants?

Yes, underground wiring must be installed in rigid metal conduit; when buried in 2 feet or more of earth, it may be installed in nonmetallic conduit or duct, or in the form of cable approved for that purpose. Where cable is used, it must be enclosed in rigid metal conduit from the point of lowest buried cable level to the point of connection to the aboveground raceway (see NEC, Article 515-5).

14-62 What are finishing-process locations?

Locations where paints, lacquers, or other flammable finishes are used regularly or are frequently applied by spraying, dipping, brushing, or other means, and where readily ignitible deposits or residues from such paints, lacquers, or finishes may occur (see NEC, Article 516-1).

14-63 What is the classification of finishing-process locations?

They are considered as Class I, Division 1 locations (see NEC, Article 516-2).

14-64 Can a direct-drive fan be used for ventilation of finishing-process areas?

No, even though an explosion-proof motor is used, because the air passing over the motor contains particles of lacquers and paints. These particles will settle out on the motor, and the exterior temperature of the motor may reach a point at which these residues would become easily ignited (see NEC, Article 516-3).

SPECIAL OCCUPANCIES AND HAZARDOUS LOCATIONS

14-65 Is there a lighting fixture approved for direct hanging in a spray-painting booth?

None have been approved, because of the low flash point of residues that collect on the glass. Illumination through panels of glass or other translucent or transparent materials is permissible only where fixed lighting units are used, the panel effectively isolates the hazardous area from the area in which the lighting units are located, the lighting units are approved for their specific locations, the panel is of material or is so protected that breakage will be unlikely, and if the arrangement is such that normal accumulations of hazardous residues on the surface of the panel will not be raised to a dangerous temperature by radiation or conduction from the illumination source (see NEC, Article 516-3).

Article 517 of the NEC was formerly termed, **Flammable Anesthetics** but is now called, **Health Care Facilities.** This broadens the coverage and give badly needed safety features.

14-66 What is a Critical Branch?

Critical Branch—A sub-system of the emergency system consisting of feeders and branch circuits supplying energy to task illumination and selected receptacles serving areas and functions related to patient care, and which can be connected to alternate power sources by one or more transfer switches. (Definition in Section 517-2 of the NEC).

14-72 What is a life support branch?

The life support branch of the emergency system supplies power centers in electrically susceptible patient locations. (definition in Section 517-2 of the NEC).

14-73 What is a reference ground point?

The terminal grounding bus which serves as the single focus for grounding the electrical equipment connected to an individual patient, or for grounding the metal or conductive furniture or other equipment within reach of the patient or a person who may be touching him. Definition in Section 517-2 of the NEC).

14-74 What is a room bonding point?

The terminal grounding bus which serves as a single focus for grounding the patient reference grounding buses and all other

Special Occupancies and Hazardous Locations

metal or conductive furniture, equipment, or structural surfaces in the room.

The bus may be located in or outside the room. The room reference grounding bus and the patient reference grounding bus may be a common bus if there is only one patient grounding bus in the room. (Definition in Section 517-2 of the NEC).

14-75 What is the classification of a flammable anesthetics storage room?

It is considered as a Class I, Division 1 location throughout the entire area. (See NEC Section 517-60a).

14-76 What classification is given to an anesthetizing location?

It is considered as a Class I, Division Location and this area extends upwards to a height of 5 feet. (See NEC, Section 517-60(a)(2)).

14-77 What type of wiring is required in a flammable-anesthetics location?

Any wiring operating at more than 8 volts between conductors must conform to Class I, Division 1, location wiring specifications and must be approved for the hazardous atmospheres involved. (See NEC Section 517-61a).

14-78 What are considered as the wiring boundaries in a flammable-anesthetics location?

Where masonry constitutes a boundary in the hazardous area, raceways embedded in the masonry are considered within the boundary itself, but any portion of the raceway located in a hollow wall space is considered to be within the hazardous area. (See NEC Section 517-61).

14-79 What special precautions pertain to circuits in anesthetizing areas?

All circuits must be supplied by approved isolating transformers, and proper ground-detector systems approved for the location must be used. (See the NEC, Section 517-63).

CHAPTER 15

Grounding and Ground Testing

15-1 What is the resistance, in general, of a continuous metallic underground water-piping system?
It is usually 3 ohms or less to ground (see NEC, Article 250-84).

15-2 What is the maximum allowable resistance of made electrodes (grounding)?
The resistance to ground cannot exceed 25 ohms (see NEC, Article 250-84).

15-3 If the resistance exceeds 25 ohms, how may this condition be corrected?
By connecting two or more electrodes in parallel (see NEC, Article 250-84).

15-4 When driving a ground rod, or made electrode, is the resistance near the foundation more or less than the resistance of a rod driven a few feet away from the foundation?
The resistance will usually be more when a rod is driven near the foundation; this is due to the fact that when the rod is driven several feet away, the earth is in a circle around the ground rod instead of in a semicircle, as it would be when driven close to the foundation.

15-5 Is the ground resistance more or less when a rod is driven into undisturbed soil?
When a rod is driven into undisturbed soil, the pressure is greater against the rod; this pressure lowers the ground resistance.

15-6 Must all driven grounds be tested?

GROUNDING AND GROUND TESTING

Since made electrodes must, where practical, have a resistance to ground not in excess of 25 ohms, the testing of ground resistance can be required by the inspection authority.

15-7 Is the common ohmmeter (DC-type) a good instrument to use when testing ground resistance?

No.

15-8 When using a common ohmmeter for ground testing, why can the results not be relied on?

Stray AC ground currents will probably be encountered. Some DC currents may also be found in the ground; these are due to electrolysis—the battery action between the moist earth and metals in contact with it, including the grounding electrode.

15-9 What instruments should be used for testing ground resistance?

A standard ground-testing meger, a battery-operated vibrator-type ground tester, or a transistor-oscillator-type ground tester. The resistance may also be tested with an isolating transformer, in conjuction with a voltmeter and an ammeter.

15-10 When testing the ground resistance on a ground rod, should the connection to the service be disconnected from the rod?

Yes.

15-11 Why should the connection to the ground rod be removed while testing?

Because of the danger of feedback into any other ground rods, equipment, etc., which would produce an inaccurate reading.

15-12 Sketch the procedure for ground-rod testing (see Fig. 1).

15-12a Why is the middle electrode (B) in Fig. 1 set at 62% of the distance from A to C?

Referring to Fig. 1a you will see that the knee of the resistance curve is at "X" or approximately 62%, and this point gives us the most accurate reading.

15-13 Sketch the procedure for testing ground resistance by the use of an isolating transformer, ammeter, and voltmeter (see Fig. 2).

GROUNDING AND GROUND TESTING

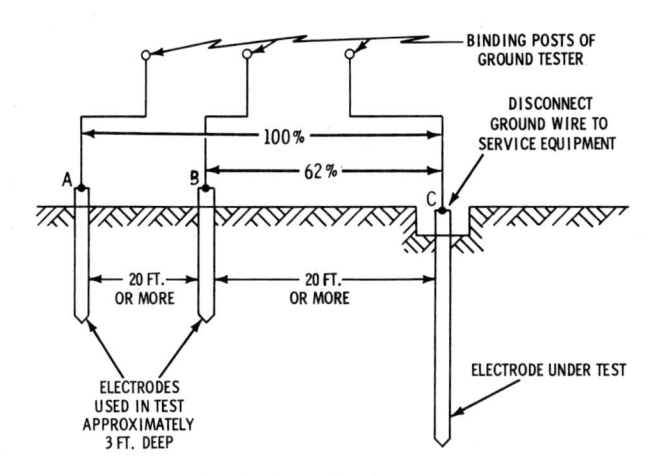

Fig. 1. Ground-rod testing.

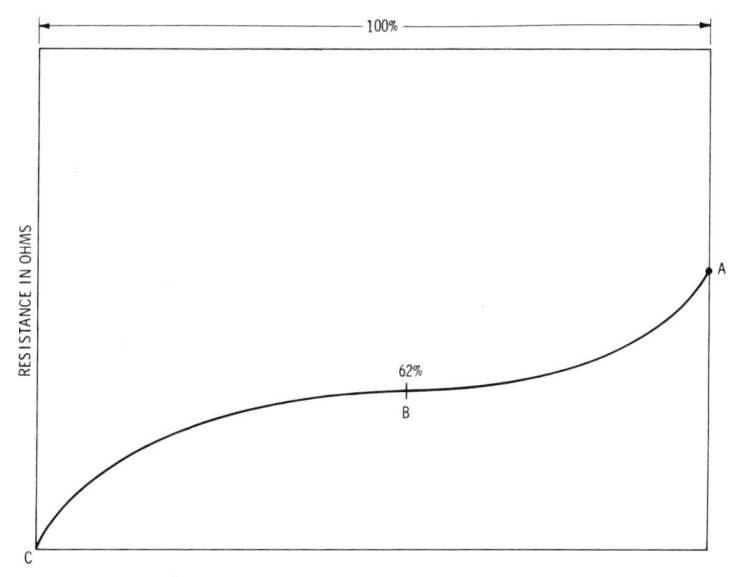

Fig. 1A Distance of B from C measured electrode.

15-14 What other method may be used to lower the resistance of a ground rod besides paralleling rods?

The use of chemicals, such as magnesium sulfate, copper sulfate, or rock salt.

GROUNDING AND GROUND TESTING

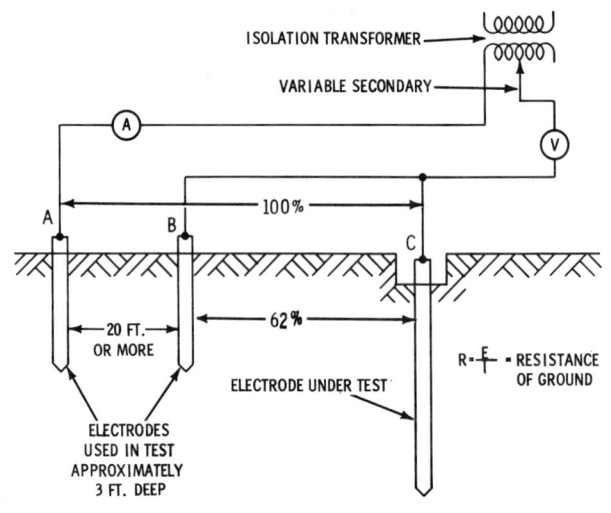

Fig. 2. Testing the ground resistance with an isolation transformer, ammeter, and voltmeter.

15-15 Approximately what quantity of chemicals is necessary for the treatment of soil to lower its ground resistance?

The first treatment should use from 40 to 90 pounds.

15-16 How is moisture added to the chemical treatment?

The normal amount of rainfall, in most places, will provide sufficient moisture. In extremely arid areas, the problem will be different.

15-17 Sketch two common methods of adding chemicals to the ground (see Fig. 3).

15-18 When paralleling ground rods, will the resistance be cut in proportion to the number of rods that are paralleled?

The resistance will not be cut proportionally. For instance, two rods paralleled and spaced 5 feet apart will cut the resistance to approximately 65% of the original resistance. Three rods in parallel and spaced 5 feet apart will cut the original resistance to approximately 42%; four rods in parallel and spaced 5 feet apart will cut the original resistance to approximately 30%.

15-19 When testing the insulation resistance, as required

GROUNDING AND GROUND TESTING

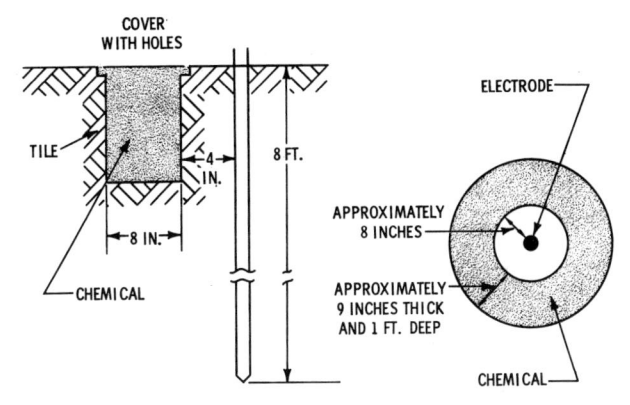

Fig. 3. Adding chemicals to the soil to lower its ground resistance.

under Article 110-20, which instrument will give the better results, an ohmmeter or a meger?

The meger will give the better results, because it produces a higher output voltage, which often shows up defects that the low-voltage ohmmeter cannot indicate.

15-20 What type of equipment may be used in and around swimming pools?

All electrical equipment must be approved for this type of use (see NEC, Article 680-2).

15-21 What is the maximum voltage allowed for underwater lighting in a swimming pool?

No lighting can be operated at more than 150 volts (see NEC, Article 680-20a(3).

15-22 May ground-fault circuit interrupters be used on underwater lighting?

Yes, (see NEC, Article 680-4e and 680-5).

15-23 May load-size conductors from ground-fault interrupters or transformers that supply underwater lighting be run with other electrical wiring or equipment?

No, they shall be kept independent. (See NEC, Article 680-5b.)

15-25 Must the noncurrent-carrying parts (metal) of lighting fixtures be grounded for the underwater lighting of swimming pools?

They must be grounded, whether exposed or enclosed in nonconducting materials (see NEC, Section 680-24 and 680-25).

Grounding and Ground Testing

15-26 How must underwater fixtures be installed in swimming pools?

Only approved fixtures may be installed, and they must be installed in the outside walls of the pool in closed recesses that are adequately drained and accessible for maintenance (see NEC, Section 680-20).

15-27 Do fixtures and fixture housings for underwater lighting have to be approved?

Yes, check in Underwriters Laboratories green book. (See NEC, Section 680-20).

15-28 May galvanized rigid conduit be used to supply Wet-Niche fixtures?

No, conduit of brass or other corrosion-resistant metal shall be used. (See NEC, Section 680-20b.)

15-29 May ordinary isolation transformers be used for supplying underwater lighting in swimming pools?

No; the transformer and enclosure must be approved for this purpose (see NEC, Section 680-5a).

15-30 How can approved equipment for swimming-pool usage be properly identified?

Look for the UL label; then look up the number and manufacturer's name in the UL listing books to ascertain whether it is listed for this type of use.

15-31 How close to a swimming pool may an attachment plug receptable be installed?

Not closer than 10 feet from the inside walls of the swimming pool (see NEC, Article 680-6).

15-32 Of what material must junction boxes that supply underwater pool lights be made?

Brass or other suitable copper alloy, if less than 4 feet from the pool perimeter and less than 8 inches above the ground or concrete (see NEC, Section 680-21a).

15-33 How high or far from the pool shall approved transformers be mounted?

They are to be located not less than 8 inches, measured from the inside bottom of the enclosure to the ground level, pool deck, or maximum pool water level, whichever provides the

GROUNDING AND GROUND TESTING

greatest elevation; also not less than 4 feet from the inside wall of the pool unless separated from the pool by a solid fence, wall or other permanent barrier. (See NEC, Section 680-21)

15-34 What type of metal equipment must be grounded in pool areas?

All metallic conduit, piping systems, pool reinforcing steel, lighting fixtures, etc., must be bonded together and grounded to a common ground; this includes all metal parts of ladders, diving boards, and their supports (see NEC, Section 680-24).

15-35 What is considered adequate in the bonding of the reinforcing bar in concrete?

Tying the reinforcing bar together with wire, as is customarily done, is considered adequate, provided the job is well done; welding, of course, would also be acceptable. (See NEC, Section 680-22a, Exception No. 1.)

15-36 May the electrical equipment be grounded to a separate grounding electrode on a swimming-pool installation?

No; the grounding must be common to the deck box or transformer ground (see NEC, Section 680-25).

15-37 What is the minimum size of grounding conductor permitted for deck boxes on swimming pools?

The grounding wire must be No. 12 AWG or larger (see NEC, Section 680-25(d)).

15-39 May metal raceways be relied on for grounding in swimming-pool areas?

No; an insulated grounding conductor is required (see NEC, Section 680-25(d)).

15-40 How must pumps, water-treating equipment, etc., in swimming-pool areas be grounded?

They shall be bonded by a solid copper conductor not smaller than No. 8 AWG. (See NEC, Section 680-22a).

15-41 May structural reinforcing steel be used for bonding nonelectrical parts in swimming-pool areas?

Yes, where the requirements of Section 250-113 are met and reliable and approved connections can be made (see NEC, Section 680-22a Exception 2).

15-42 What are the clearances for service-drop conductors

273

GROUNDING AND GROUND TESTING

in swimming-pool locations?

They must be installed not less than 10 feet horizontally from the pool edge, diving structures, observation stands, towers, or platforms, and must not be installed above the swimming pool or surrounding area within the 10-foot area (see NEC, Section 680-8).

15-43 What type of wiring is required in theaters and assembly halls?

The wiring must be metal raceways, or Type ALS, MC or MI cable, with some exceptions (see NEC, Section 518-3).

15-44 Where the assembly area is less than a 100-person capacity, may the type of wiring be altered?

Yes, Type AC metal-clad cable, knob-and-tube, or Type NM cable may be used. However, it is not considered good practice to use anything other than metal raceways. Also, most local ordinances or laws prohibit the use of anything other than metal raceways in public places of assembly.

15-45 How may the population capacity be determined in places of assembly?

Refer to the NFPA Life Safety Code (No. 101) (Table 1, page 276).

The maximum number of persons in a standing position to occupy floor areas, such as stairway landings or areas of temporary refuge, shall be computed at 3 square feet per person. The NFPA Life Safety Code No. 101 should ideally be a part of every electrician's library.

15-46 In the following diagram (Fig. 4) of a 120/240-volt, three-wire circuit, load A is 10 amperes and load B is 5 amperes. Will the neutral carry any current, and if so, how much will it carry?

(a) No current
(b) 10 amperes
(c) 15 amperes
(d) *5 amperes*

**15-47 In the following 120/240-volt, three-wire circuit (Fig. 5), the neutral is open at point X. The resistance of load A is 10 ohms; load B, 12 ohms; load C, 24 ohms; and load D, 20 ohms. What is the voltage drop of loads B

GROUNDING AND GROUND TESTING

and C, respectively?
(a) B = 160 volts; C = 80 volts
(b) *B = 80 volts; C = 160 volts*
(c) B = 120 volts; C = 120 volts
(d) B = 240 volts; C = 240 volts

15-48 **If the neutral is No. 10 wire or larger, what appliance, if any, may use the neutral as an equipment ground?**
(a) None
(b) Grounding-type receptacle
(c) Electric motor
(d) *Electric range*

15-49 **When pulling current-carrying conductors into conduits, at what number of such conductors does derating of conductor current-carrying capacity begin?**
(a) *Four*
(b) Six
(c) No derating
(d) Three

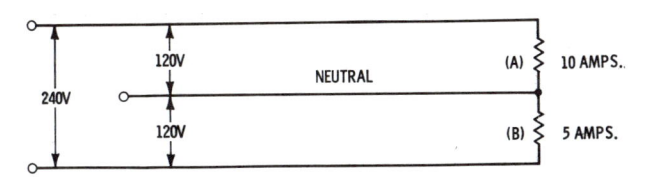

Fig. 4. The neutral lead carries 5 amperes of current.

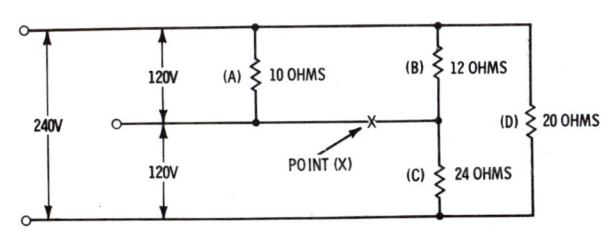

Fig. 5. With the neutral lead open at point X, the voltage drop of load B is 80 volts, and the voltage drop of load C is 160 volts.

GROUNDING AND GROUND TESTING

Table 1.

Occupancy	Sq. ft. per person
Places of Assembly	15 net
Areas of concentrated use without fixed seating	7 net
Standing space	3 net
Store, street floor, and sales basement	30 gross
Other floors	60 gross
Storage, shipping	100 gross
Educational occupancies	
Classroom area	20 net
Shops and other vocational areas	50 net
Office, factory, and workroom	100 gross
Hotel and apartment	200 gross
Institutional	
Sleeping departments	120 gross
In-patient departments	240 gross

15-50 Under what condition, if any, is it possible or permissible to put overload protection in a neutral conductor?
  (a) Always permissible
  (b) Never permissible
  (c) *Where the overload device simultaneously opens all conductors*

15-51 Under what conditions may a yellow neutral be used?
  (a) When extra color coding is advantageous
  (b) *Never*
  (c) On rewire work

15-52 The total load on a circuit for air conditioning only cannot exceed what percentage of the branch-circuit load?
  (a) 115%
  (b) 75%
  (c) *80%* (see NEC, Section 440-62(b)).

15-53 The total load of an air conditioner cannot exceed what percentage of the branch-circuit load on a circuit that also supplies lighting?
  (a) *50%* (see NEC, Section 440-62(c)).

Grounding and Ground Testing

(b) 75%
(c) 80%

15-54 Hazardous concentrations of gases or vapors are not normally present in Class I, Division 1 locations.
T *F* (see NEC, Article 500-4a)

15-55 A location where cloth is woven should be designated as Class III, Division 1.
T F (see NEC, Article 500-6a)

15-56 Resistance- and reactive-type dimmers may be placed in the ungrounded conductor.
T F (see NEC, Article 520-25b)

15-57 With outline lighting, stranded conductors need not be soldered when pin-type sockets are used.
T F (see NEC, Article 600-21e)

15-58 A separate connection from the service drop is never acceptable as an emergency service installation.
T *F* (see NEC, Article 700-10)

15-59 Portable, explosion-proof lighting units may be employed inside a spray booth during operation.
T *F* (see NEC, Article 516-3)

15-60 Askarel-filled transformers may be used in Class II, Division 2 locations
T F (see NEC, Article 502-2b-2)

15-62 A plug receptable exclusively for the janitor's use may not be tapped from the emergency circuit wiring.
T F (see NEC, Article 700-13)

15-63 Type NM (nonmetallic) cable may be used for under-plaster extensions.
T *F* (see NEC, Article 336-3a)

15-64 Heating cable may not be used with metal lath.
T *F* (see NEC, Article 424-41)

15-65 Time switches need not be of the externally operated type.
T F (see NEC, Article 380-5)

15-66 Heating cables in a concrete floor must be placed on at least 2-inch centers.
T *F* (see NEC, Section 426-24)

15-67 Double-throw knife switches may be mounted either

GROUNDING AND GROUND TESTING

vertically or horizontally.
T F   (see NEC, Article 380-6)

15-68 Outlets for heavy-duty lampholders must be rated at a minimum load of
(a) *600 volt-amperes* (see NEC Section 220-2c)
(b) 1320 volt-amperes
(c) 2 amperes
(d) 5 amperes

15-69 Where permissible, the demand factor applied to that portion of the unbalanced neutral load in excess of 200 amperes is
(a) 40%
(b) 80%
(c) *70%* (see NEC, Section 220-22)
(d) 60%

15-70 The largest standard cartridge fuse rating, in amperes, is
(a) *6000* (see NEC, Section 240-61)
(b) 1200
(c) 1000
(d) 600

15-72 Fuses must never be connected in multiple.
T F   (see NEC, Article 240-18)

15-73 No. 18 copper wire may be employed for grounding a portable device used on a 20-ampere circuit.
T F   (see NEC, Article 250-95)

15-74 Secondary circuits of wound-rotor induction motors require overcurrent protection.
T *F*   (see NEC, Article 430-32d)

15-75 A neutral feeder conductor is sometimes smaller than the ungrounded conductors.
T F   (see NEC, Article 220-22)

15-76 The largest size conductor permitted in underfloor raceways is
(a) *largest that raceway is designed for* (see NEC, Article 354-4)
(b) No. 0
(c) 250 MCM
(d) No. 000

GROUNDING AND GROUND TESTING

15-77 The maximum rating of a portable appliance used on a 20-ampere branch circuit must be
(a) *16 amperes* (see NEC, Article 210-23)
(b) 10 amperes
(c) 20 amperes
(d) 12 amperes

15-78 The demand factor that may be applied to the neutral of an electric range is
(a) 40%
(b) 80%
(c) *70%* (see NEC, Article 220-22)
(d) 60%

15-79 The unprotected length of a tap conductor having a current-carrying capacity equal to one-third that of the main conductor must be no greater than
(a) 15 feet
(b) *25 feet* (see NEC, Article 240-21, Exception No. 3)
(c) 20 feet
(d) 10 feet

15-80 The minimum feeder allowance for show-window lighting expressed in watts per linear foot shall be
(a) 100 watts
(b) *200 watts* (see NEC, Article 220-12)
(c) 300 watts
(d) 500 watts

AUDEL BOOKS practical reading for profit

APPLIANCES

Air Conditioning   (23159)

Domestic, commercial, and automobile air conditioning fully explained in easily-understood language. Troubleshooting charts aid in making diagnosis and repair of system troubles.

Commercial Refrigeration   (23195)

Installation, operation, and repair of commercial refrigeration systems. Included are ice-making plants, locker plants, grocery and supermarket refrigerated display cases, etc. Trouble charts aid in the diagnosis and repair of defective systems.

Air Conditioning and Refrigeration Library—2 Vols.   (23196)

Home Appliance Servicing—3rd Edition   (23214)

A practical "How-To-Do-It" book for electric & gas servicemen, mechanics & dealers. Covers principles, servicing and repairing of home appliances. Tells how to locate troubles, make repairs, reassemble and connect, wiring diagrams and testing methods. Tells how to fix electric refrigerators, washers, ranges, toasters, ironers, broilers, dryers, vacuum sweepers, fans, and other appliances.

Home Refrigeration and Air Conditioning (23133)

Covers basic principles, servicing, operation, and repair of modern household refrigerators and air conditioners. Automotive air conditioners are also included. Troubleshooting charts aid in trouble diagnosis. **A gold mine of essential facts for engineers, servicemen, and users.**

Oil Burners   (23151)

Provides complete information on all types of oil burners and associated equipment. Discusses burners—blowers—ignition transformers—electrodes—nozzles—fuel pumps—filters—controls. Installation and maintenance are stressed. Troubleshooting charts permit rapid diagnosis of system troubles and possible remedies to correct them.

AUTOMOTIVE

Automobile Guide   (23192)

New revised edition. Practical reference for auto mechanics, servicemen, trainees, and owners. Explains theory, construction, and servicing of modern domestic motorcars. FEATURES: All parts of an automobile—engines—pistons—rings—connecting rods—crankshafts—valves—cams—timing—cooling systems—Fuel-feed systems—carbureators — automatic choke — transmissions — clutches — universals — propeller shafts—dierentials—rear axles—running gear—brakes—wheel alignment—steering gear—tires—lubrication—ignition systems—generators and alternators—starters—lighting systems—batteries—air conditioning—cruise controls—emission control systems.

Auto Engine Tune-up   (23181)

New revised edition. This popular how-to-do-it guide shows exactly how to tune your car engine for extra power, gas economy, and fewer costly repairs. New emission-control systems are explained along with the proper methods for correcting faults and making adjustments to keep these systems in top operating condition.

Automotive Library—2 Vols.   (23198)

Diesel Engine Manual   (23199)

A **practical treatise on the theory, operation and maintenance of modern Diesel engines.** Explains Diesel principles—valves—timing—fuel pumps—pistons and rings—cylinders—lubrication—cooling system—fuel oil—engine indicator—governors—engine reversing—answers on operation—calculations. AN IMPORTANT GUIDE FOR ENGINEERS, OPERATORS, STUDENTS.

Gas Engine Manual   (23061)

A **completely practical book covering the construction, operation and repair of all types of modern gas engines.** Part I covers gas-engine principles; engine parts; auxiliaries; timing methods; ignition systems. Part II covers troubleshootng, adjustment and repairs.

BUILDING AND MAINTENANCE

Answers on Blueprint Reading   (23041)

Covers all types of blueprint reading for mechanics and builders. The man who can read blueprints is in line for a better job. This book gives you the secret language, step by step in easy stages. NO OTHER TRADE BOOK LIKE IT.

Building Construction and Design   (23180)

A completely revised and rewritten version of Audel's **Architects and Builders Guide.** New illustrations and extended coverage of material makes this treatment of the subject more valuable than ever. Anyone connected in any way with the building industry will profit from the information contained in this book.

Building Maintenance   (23140)

A **comprehensive book on the practical aspects of building maintenance.** Chapters are included on: painting and decorating; plumbing and pipe fitting; carpentry; calking and glazing; concrete and masonry; roofing; sheet metal; electrical maintenance; air conditioning and refrigeration; insect and rodent control; heating maintenance management; cutodial practices: A BOOK FOR BUILDING OWNERS, MANAGERS, AND MAINTENANCE PERSONNEL.

Gardening & Landscaping   (23229)

A comprehensive guide for the homeowner, industrial, municipal, and estate groundskeepers. Information on proper care of annual and perennial flowers; various house plants; greenhouse design and construction; insect and rodent control; complete lawn care; shrubs and trees; and maintenance of walks, roads, and traffic areas. Various types of maintenance equipment are also discussed.

Carpenters & Builders Library—4 Vols.   (23169)

A **practical illustrated trade assistant on modern construction for carpenters, builders, and all woodworkers.** Explains in practical, concise language and illustrations all the principles, advances and short cuts based on modern practice. How to calculate various jobs.

Vol. 1—(23170)—Tools, steel square, saw filing, joinery, cabinets.
Vol. 2—(23171)—Mathematics, plans, specifications, estimates.
Vol. 3—(23172)—House and roof framing, laying out, foundations.
Vol. 4—(23173)—Doors, windows, stairs, millwork, painting.

Carpentry and Building   (23142)

Answers to the problems encountered in today's building trades. The actual questions asked of an architect by carpenters and builders are answered in this book. No apprentice or journeyman carpenter should be without the help this book can offer.

Do-It-Yourself Encyclopedia   (23207)

An **all-in-one home repair and project guide for all do-it-yourselfers.** Packed with step-by-step plans, thousands of photos, helpful charts. A really authentic, truly monumental, home-repair and home-project guide.

Home Workshop & Tool Handy Book   (23208)

The most modern, up-to-date manual ever designed for home craftsmen and do-it-yourselfers. Tells how to set up your own home workshop, (basement, garage, or spare room), all about the various hand and power tools (when, where, and how to use them, etc.). Covers both wood- and metal-working principles and practices. An all-in-one workshop guide for handy men, professionals and students.

Plumbers and Pipe Fitters Library—3 Vols.   (23155)

A practical illustrated trade assistant and reference for master plumbers, journeyman and apprentice pipe fitters, gas fitters and helpers, builders, contractors, and engineers. Explains in simple language, illustrations, diagrams, charts, graphs and pictures, the principles of modern plumbing and pipe-fitting practices.

Vol. 1—(23152)—Materials, tools, calculations.
Vol. 2—(23153)—Drainage, fittings, fixtures.
Vol. 3—(23154)—Installation, heating, welding

Masons and Builders Library—2 Vols.   (23185)

A practical illustrated trade assistant on modern construction for bricklayers, stonemasons, cement workers, plasterers, and tile setters. Explains in clear language and with detailed illustrations all the principles, advances, and shortcuts based on modern practice—including how to figure and calculate various jobs.

Vol. 1—(23182)—Concrete, Block, Tile, Terrazzo.
Vol. 2—(23183)—Bricklaying, Plastering, Rock Masonry, Clay Tile.

Upholstering   (23189)

Upholstering is explained for the average householder and apprentice upholsterer in this Audel text. Selection of coverings, stuffings, springs, and other upholstering material is made simple. From repairing and regluing of the bare frame, to the final sewing or tacking, for antiques and most modern pieces, this book gives complete and clearly written instructions and numerous illustrations.

Questions and Answers for Plumbers Examinations   (23206)

Many questions are answered as to types of fixtures to use, size of pipe to install, design of systems, size and location of septic tank systems, and procedures used in installing material. Subjects such as traps, cleanouts, drainage, vents, water supply and distribution, sewage treatment, plastic pipe, steam and hot water fittings, just to name a few.

Wood Furniture, Finishing, Refinishing Repairing   (23216)

The basic technical and practical information needed for complete wood finishing. This book presents the fundamentals of furniture repair, both veneer and solid wood and complete refinishing procedures, which includes stripping the old finish, sanding, selecting the finish and using wood fillers. Complete step-by-step procedure for antiquing, painting, staining, flocking, inlay patterns and gold- and silver-leaf finishing. Various woods, along with actual grain photos, are discussed as to their characteristics and their reactions to various finishes.

Building a Vacation Home: Step-by-Step   (23222)

Step-by-step procedure in the construction of this vacation home includes fifteen chapters, numerous illustrations, actual photographs, and a complete set of blueprints, showing cabinet construction as well as building and foundation structure. It points out difficulties which might be avoided by informing you of the proper procedures for obtaining property lines, building and sewage permits, and various hidden costs which are bound to occur.

ELECTRICITY-ELECTRONICS

Wiring Diagrams for Light & Power   (23232)

Electricians, wiremen, linemen, plant superintendents, construction engineers, electrical contractors, and students will find these diagrams a valuable source of practical help. Each diagram is complete and self-explaining. A PRACTICAL HANDY BOOK OF ELECTRICAL HOOK-UPS.

Electric Motors   (23150)

Covers the construction, theory of operation, connection, control, maintenance, and troubleshooting of all types of electric motors. A HANDY GUIDE FOR ELECTRICIANS AND ALL ELECTRICAL WORKERS.

Practical Electricity   (23218)

This updated version is a ready reference book, giving complete instruction and practical information on the rules and laws of electricity—maintenance of electrical machinery—AC and DC motors—wiring diagrams—lighting—house and power wiring—meter and instrument connections—transformer connectors—circuit breakers—power stations—automatic substations. THE KEY TO A PRACTICAL UNDERSTANDING OF ELECTRICITY.

House Wiring—3rd Edition   (23224)

Answers many questions in plain simple language concerning all phases of house wiring. A ready reference book with over 100 illustrations and concise interpretations of many rulings contained in the National Electrical Code. Electrical contractors, wiremen, and electricians will find this book invaluable as a tool in the electrical field

Guide to the 1975 National Electric Code   (23223)

This important and informative book is now revised to conform to the 1975 National Electrical Code. Offers an interpretation and simplification of the rulings contained in the Code. Electrical contractors, wiremen, and electricians will find this book invaluable for a more complete understanding of the NEC.

Questions and Answers for Electricians Examinations   (23225)

Newly revised to conform to the 1975 National Electrical Code. A practical book to help you prepare for all grades of electricians examinations. A helpful review of fundamental principles underlying each question and answer needed to prepare you to solve any new or similar problem. Covers the NEC; questions and answers for license tests; Ohm's law with applied examples, hook-ups for motors, lighting, and instruments. A COMPLETE REVIEW FOR ALL ELECTRICAL WORKERS.

Electrical Library—6 Vols.   (23236)

Electric Generating Systems   (23179)

Answers many questions concerning the selection, installation, operation, and maintenance of engine-driven electric generating systems for emergency, standby, and away-from-the-power-line applications. Private homes, hospitals, radio and television stations, and pleasure boats are only a few of the installations that owners either desire or require for primary power or for standby use in case of commercial power failure. THE MOST COMPREHENSIVE COVERAGE OF THIS SUBJECT TO BE FOUND TODAY.

Electrical Course for Apprentices and Journeymen (23209)

A basic study course for apprentice or journeyman electricians which may be used as a classroom or self-taught program. This book can be utilized without any other books on electrical theory. Review questions included.

Electronic Security Systems (23205)

Protect your home and business. A basic and practical text written for the electrician, electronic technician, security director and do-it-yourself householder or businessman. Such subjects as sensors and encoders, indicators and alarms, electrical control and alarm circuits, security communications, closed circuit television, and security system installations are covered.

ENGINEERS-MECHANICS-MACHINISTS

Machinists Library (23174)

Covers modern machine-shop practice. Tells how to set up and operate lathes, screw and milling machines, shapers, drill presses and all other machine tools. A complete reference library. A SHOP COMPANION THAT ANSWERS YOUR QUESTIONS.

Vol. 1—(23175)—Basic Machine Shop.
Vol. 2—(23176)—Machine Shop.
Vol. 3—(23177)—Toolmakers Handy Book.

Millwrights and Mechanics Guide— 2nd Edition (23201)

Practical information on plant installation, operation, and maintenance for millwrights, mechanics, maintenance men, erectors, riggers, foremen, inspectors, and superintendents. Partial contents: • Drawing and Sketching • Machinery Installation • Power-Transmission Equipment • Couplings • Packing and Seals • Bearings • Structural Steel • Mechanical Fasteners • Pipe Fittings and Valves • Carpentry • Sheet-Metal Work • Blacksmithing • Rigging • Electricity • Welding • Mathematics and much more.

Practical Guide to Mechanics (23102)

A convenient reference book valuable for its practical and concise explanations of the applicable laws of physics. Presents all the basics of mechanics in everyday language, illustrated with practical examples of their applications in various fields.

Power Plant Engineers Guide—2nd Edition (23220)

A complete steam or diesel power plant engineers library. This book covers (in question-and-answer form) facts for all engineers, fireman, water tenders, oilers, and operators of steam and diesel power systems. A valuable book to the applicant for engineer's and fireman's licenses.

Questions & Answers for Engineers and Firemans Examinations (23217)

An aid for stationary, marine, diesel & hoisting engineers' examinations for all grades of licenses. A new concise review explaining in detail the principles, facts and figures of practical engineering. Questions & Answers.

Welders Guide—2nd Edition (23202)

New revised edition. Covers principles of electric, oxyacetylene, thermit, unionmelt welding for sheet metal; spot and pipe welds; pressure vessels; aluminum, copper brass, bronze, plastics, and other metals; airplane work; surface hardening and hard facing; cutting brazing; underwater welding; eye protection. EVERY WELDER SHOULD OWN THIS GUIDE.

Mechanical Trades Pocket Manual   (23215)

This "paper back—pocket size manual" provides reference material for mechanical tradesman. The handbook covers methods, tools, equipment, procedures, etc. in convenient form and plain language to aid the mechanic in performance of day-to-day tasks concerned with installation, maintenance, and repair of machinery and equipment.

FLUID POWER

Pumps   (23167)

A **detailed book on all types of pumps** from the old-fashioned kitchen variety to the most modern types. Covers construction, application, installation, and troubleshooting.

MATHEMATICS

Practical Mathematics for Everyone—2 Vols.   (23112)

A **concise and reliable guide to the understanding of practical mathematics.** People from all walks of life, young and old alike, will find the information contained in these two books just what they have been looking for. The mathematics discussed is for the everyday problems that arise in every household and business.
Vol. 1—(23110)—Basic Mathematics.
Vol. 2—(23111)—Financial Mathematics.

OUTBOARD MOTORS

Outboard Motors & Boating   (23168)

Provides the information necessary to adjust, repair, and maintain all types of outboard motors. Valuable information concerning boating rules and regulations is also included.

RADIO-TELEVISION-AUDIO

Handbook of Commercial Sound Installations   (23126)

A **practical complete guide to planning commercial systems,** selecting the most suitable equipment, and following through with the most proficient servicing methods. For technicians and the professional and businessman interested in installing a sound system.

Practical Guide to Auto Radio Repair   (23128)

A **complete servicing guide for all types of auto radios,** including hybrid, all-transistor, and FM . . . PLUS removal instructions for all late model radios. Fully illustrated.

Practical Guide to Servicing Electronic Organs   (23132)

Detailed, illustrated discussions of the **operation and servicing of electronic organs.** Including models by Allen, Baldwin, Conn, Hammond, Kinsman, Lowrey, Magnavox, Thomas, and Wurlitzer.

Radioman's Guide   (23163)

Audel best-seller, containing the latest information on radio and electronics from the basics through transistors. Covers radio fundamentals—Ohm's law—physics of sound as related to radio—radio-wave transmission—test equipment—power supplies—resistors, inductors, and capacitors—transformers—vacuum tubes—transistors—speakers—antennas—troubleshooting. A complete guide and a perfect preliminary to the study of television servicing.

Television Service Manual   (23162)

Includes the latest designs and information. Thoroughly covers television with transmitter theory, antenna designs, receiver circuit operation and the picture tube. Provides the practical information necessary for accurate diagnosis and repair of both black-and-white and color television receivers. A MUST BOOK FOR ANYONE IN TELEVISION.

Radio-TV Library—2 Vol.   (23161)

SHEET METAL

Sheet Metal Workers Handy Book   (23046)

Containing practical information and important facts and figures. Easy to understand. Fundamentals of sheet metal layout work. Clearly written in everyday language. Ready reference index.

TO ORDER AUDEL BOOKS mail this handy form to
Theo. Audel & Co., 4300 W. 62nd
Indianapolis, Indiana 46268

Please send me for FREE EXAMINATION books marked (x) below. If I decide to keep them I agree to mail $3 in 10 days on each book or set ordered and further mail ⅓ of the total purchase price 30 days later, with the balance plus shipping costs to be mailed within another 30 days. Otherwise, I will return them for refund.

APPLIANCES
- ☐ (23196) Air Conditioning & Refrigeration Library (2 Vols.) 13.75
  - ☐ (23159) Air Conditioning 6.95
  - ☐ (23195) Commercial Refrigeration 7.50
- ☐ (23214) Home Appliance Servicing 7.95
- ☐ (23133) Home Refrigeration and Air Conditioning 8.95
- ☐ (23151) Oil Burners 6.25

AUTOMOTIVE
- ☐ (23198) Automotive Library (2 Vols.) 15.95
  - ☐ (23192) Automobile Guide 10.25
  - ☐ (23181) Auto Engine Tune-Up 6.75
- ☐ (23199) Diesel Engine Manual 8.50
- ☐ (23061) Gas Engine Manual 6.50

BUILDING AND MAINTENANCE
- ☐ (23041) Answers on Blueprint Reading 6.50
- ☐ (23180) Building Construction and Design 6.75
- ☐ (23140) Building Maintenance 6.75
- ☐ (23169) Carpenters and Builders Library (4 Vols.) 21.25
  - ☐ Single Volumes sold separately ea. 5.50
- ☐ (23142) Carpentry and Building 7.50
- ☐ (23207) Do-It-Yourself Encyclopedia 13.50
- ☐ (23229) Gardening & Landscaping 7.95
- ☐ (23208) Home Workshop & Tool Handy Book 6.50
- ☐ (23185) Masons & Builders Library (2 Vols.) 12.95
  - ☐ Single Volumes sold separately ea. 6.95
- ☐ (23189) Upholstering 6.75
- ☐ (23155) Plumbers and Pipe Fitters Library (3 Vols.) 15.50
  - ☐ Single Volumes sold separately ea. 5.50
- ☐ (23206) Q & A for Plumbers Exam 6.95
- ☐ (23216) Wood Furniture Finishing, Repair 6.95
- ☐ (23222) Building a Vacation Home 7.95

ELECTRICITY-ELECTRONICS
- ☐ (23179) Electric Generating Systems 6.75
- ☐ (23236) Electrical Library (6 Vols.) 38.00
  - ☐ (23232) Wiring Diagrams for Light and Power 6.95
  - ☐ (23150) Electric Motors 6.95
  - ☐ (23218) Practical Electricity 6.95
  - ☐ (23224) House Wiring (3rd Edition) 6.50
  - ☐ (23223) Guide to the 1975 National Electrical Code 8.50
  - ☐ (23225) Questions & Answers for Electrician's Exams 6.50
- ☐ (23205) Electronic Security Systems 6.95
- ☐ (23209) Electrical Course for Apprentices & Journeymen 6.95

ENGINEERS-MECHANICS-MACHINISTS
- ☐ (23174) Machinists Library (3 Vols.) 19.50
  - ☐ Single Volumes sold separately ea. 6.75
- ☐ (23202) Welders Guide (2nd Edition) 10.95
- ☐ (23215) Mechanical Trades Pocket Manual 3.95
- ☐ (23201) Millwrights and Mechanics Guide (2nd Edition) 10.95
- ☐ (23220) Power Plant Engineers Guide 9.95
- ☐ (23102) Practical Guide to Mechanics 5.50
- ☐ (23217) Q&A for Engineers and Firemans Exams 7.95

FLUID POWER
- ☐ (23167) Pumps 8.95

MATHEMATICS
- ☐ (23112) Practical Math for Everyone (2 Vols.) 10.25
  - ☐ Single Volumes sold separately ea. 5.50

OUTBOARD MOTORS
- ☐ (23168) Outboard Motors and Boating 5.50

RADIO-TELEVISION-AUDIO
- ☐ (23126) Handbook of Commercial Sound Installations 5.95
- ☐ (23128) Practical Guide to Auto Radio Repair 4.95
- ☐ (23132) Practical Guide to Servicing Electronic Organs 5.50
- ☐ (23161) Radio & Television Library (2 Vols.) 13.75
  - ☐ (23163) Radiomans Guide 6.75
  - ☐ (23162) Television Service Manual 7.50

SHEET METAL
- ☐ (23046) Sheet Metal Workers Handy Book 4.50

Prices Subject to Change Without Notice

Name _____

Address _____

City _____ State _____ Zip _____

Occupation _____ Employed by _____

☐ **SAVE SHIPPING CHARGES! Enclose Full Payment With Coupon and We Pay Shipping Charges.** PRINTED IN USA